Nick Vandome

eBay.co.uk

Second Edition

In easy steps is an imprint of Computer Step
Southfield Road · Southam
Warwickshire CV47 0FB · United Kingdom
www.ineasysteps.com

Second Edition

Printed and bound in the United Kingdom

ISBN-13 978-1-84078-322-3
ISBN-10 1-84078-322-2

Contents

1 Introducing eBay

eBay is an online marketplace, which has managed to become a multi-million pound business while still retaining a genuine sense of being an online community where all of the members can influence the operation of the site. This chapter introduces the workings of eBay and shows how you can start using the world's largest online retail market.

History of eBay

Don't forget

eBay has five basic values by which the site operates. These can be accessed in full from the Community pages within the eBay site.

Don't forget

There are various stories relating to the first item that was sold on eBay: one says it was a broken laser pointer, while the other claims it was a Pez dispenser for peppermint sweets. The official version on eBay goes with the Pez dispenser option.

Don't forget

Some of the more unusual items put up for sale on eBay include a jar of mud from the Glastonbury Festival, a 50,000-year-old mammoth skeleton, Britney Spears' used bubble gum and a Vulcan Bomber.

Like all great inventions, simplicity is one of the foundations of eBay. It is not an online shop, as such, and it does not directly deal with goods or financial transactions. What it does do is enable buyers and sellers of a huge variety of products to do business with each other in a safe, friendly and efficient online environment. Essentially, eBay is the online middleman that facilitates the transactions without being involved in them directly.

eBay began in California in 1995 when Pierre Omidyar hit on a simple but brilliant idea for an online business: to create a trading environment where buyers and sellers could meet to sell products. eBay was launched in September 1995, at the height of the dotcom boom. Originally, Pierre had wanted to call it Echo Bay, but as that name was already taken by another business it had to be shortened to the now familiar global brand, eBay.

The growth of eBay has been astonishing: it has expanded to the point where it has over 181 million registered users worldwide and sites in over 33 countries around the world. On eBay.co.uk there are over 10 million registered users.

The original idea of eBay was that it would be for individuals to sell items through auctions: the item would be placed on the site and the highest bid would secure it. This is still the essence of eBay, although two important developments have taken place since the site was first launched. Firstly, individuals and mainstream retailers have set up full-time businesses through the website. Secondly, more and more high-priced items, such as cars and, in some countries, houses, are now available on the site.

Although there are strict rules about what can be sold on eBay, almost anything can be put up for auction as long as it meets these rules. If you are looking to buy and sell products in a wide variety of categories then you cannot afford to ignore eBay: it is a huge global marketplace, it is largely self-regulating and it is great fun.

What you can do on eBay

Although eBay is constantly evolving and developing, the essence of the site is still the same: a seller posts an item for auction, people bid for it, and the highest bidder purchases the item. While buying and selling remains the cornerstone of the eBay community, there is a lot more that can be done:

Browse

You can browse through the massive eBay database of items currently for sale. This is not only useful for buyers, it is also an invaluable way for sellers to see if there are similar items to theirs for sale.

View auctions

It is perfectly possible and allowable to view auctions that are currently taking place, without having to place a bid. This gives you a good idea of how the process works and can be a good indicator of the perceived value of certain items.

Buy It Now

If you do not want to participate in an auction, it is still possible to buy some items at a single click, with the Buy It Now button. This is an option that allows sellers to enter the product at a fixed price as well as, or instead of, the auction.

Feedback

Feedback is vital to the smooth running of eBay and both buyers and sellers are actively encouraged to leave feedback whenever they have been involved in any sort of transaction. Once you have bought or sold a few items you will come to realise the importance of feedback.

Customise

How you use eBay can be customised using the My eBay page that is available to every registered user.

Communicate

A vital part of any auction is the facility to ask the seller questions about the product they are selling. In addition, there is also a vibrant eBay online community where all issues relating to eBay, and other topics, can be discussed.

Beware

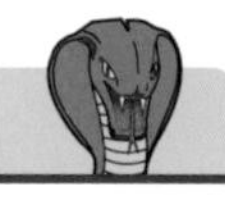

Children under the age of 18 are not allowed to register as members on eBay, but they are permitted to use the site under the supervision of an adult.

Don't forget

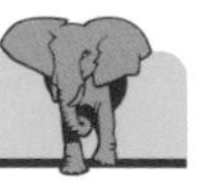

For more information on the Buy It Now option, see Chapter Four.

Don't forget

For more information on Feedback, see Chapter Nine.

Don't forget

For more information on the My eBay page, see Chapter Three.

First view

When the eBay site is first accessed the opening screen is, in effect, an introduction to eBay and, while some of the functions can be accessed, you have to register before you can start buying and selling items.

To view the initial eBay home page, enter **www.ebay.co.uk** into your browser:

1. The eBay home page contains links to all areas of the site

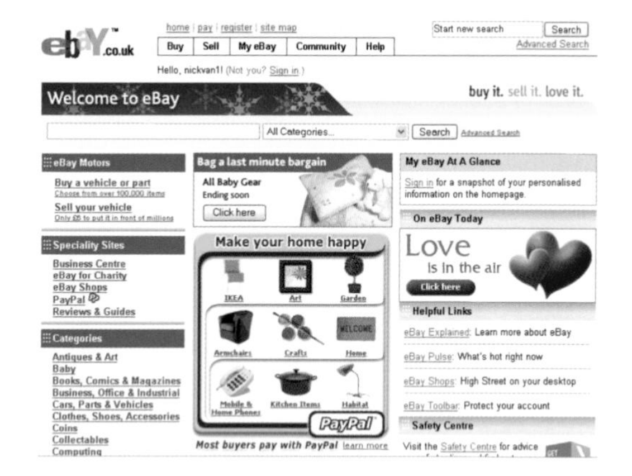

In some previous versions of eBay, and for some global sites, the initial page looks slightly different depending on whether you have registered.

2. Click one of the options on the main navigation bar to move to that page

home | pay | register | site map

Buy	Sell	My eBay	Community	Help

3. Certain areas within eBay can be viewed even if you have not registered. Here, you can see the categories of goods that are for sale

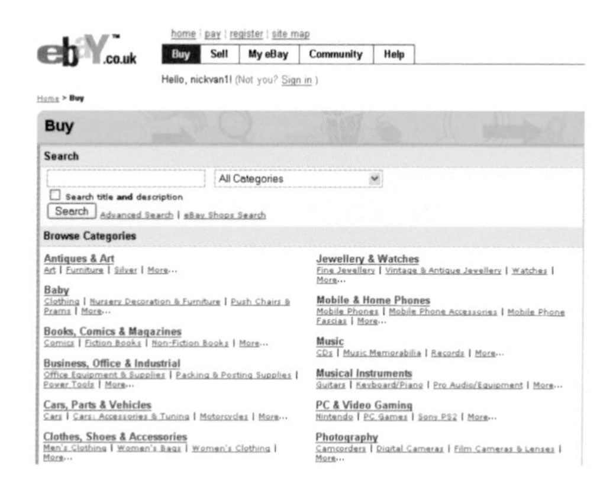

Getting registered

If you want to do anything more on eBay than have a general look around then you have to become a registered user. This is free, and once you are registered you can start buying and selling items on eBay. Since eBay takes issues of security and personal information very seriously, you will, in most cases, have to enter your credit card details to confirm that the name and address you have given is correct. To register on eBay:

1. Click here on the home page

register

2. Enter your personal details, such as your name, address and telephone number

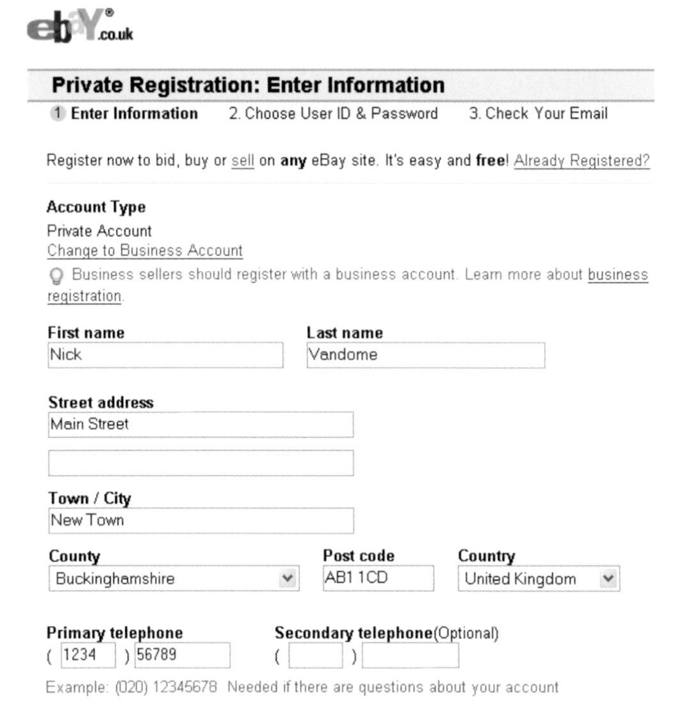

ebay.co.uk

Private Registration: Enter Information

1 Enter Information 2. Choose User ID & Password 3. Check Your Email

Register now to bid, buy or sell on **any** eBay site. It's easy and **free**! Already Registered?

Account Type
Private Account
Change to Business Account
Business sellers should register with a business account. Learn more about business registration.

First name Nick
Last name Vandome

Street address Main Street

Town / City New Town

County Buckinghamshire
Post code AB1 1CD
Country United Kingdom

Primary telephone (1234) 56789
Secondary telephone(Optional) ( )

Example: (020) 12345678 Needed if there are questions about your account

Don't forget

You can view items for sale without having to register on eBay. However, if you want to place a bid, or find out more information about an auction, the seller or other bidders, you have to go through the registration process.

...cont'd

3. Enter the relevant registration details including email address, User ID and a password

Date of Birth
01 | January | --Year-- 1965
You must be at least 18 years old to use eBay

Important: A working email address is required to complete registration.

Email address
nickvandome@hotmail.com
Examples: myname@aol.com or myname@yahoo.com

Re-enter email address
nickvandome@hotmail.com

eBay's User Agreement and Privacy Policy
In order to use eBay, you must first read and agree to eBay's User Agreement. These are the terms and conditions that apply to your use of eBay. eBay's User Agreement can be viewed and printed by clicking here.

You should also read eBay's Privacy Policy to learn about the ways in which we use and protect your personal information. The Privacy Policy can be viewed and printed by clicking here.

By checking the boxes below, I confirm the following:
- [x] I have read and accept the User Agreement and I have read the Privacy Policy.
- [x] I agree to receive communications from eBay and understand that I can change my notification preferences at any time in My eBay.
- [x] I must be an adult (18 years old) to trade on eBay and I certify that I am an adult and can enter into this Agreement.

Continue >

4. Click Continue

5. If you have entered any incorrect details, or missed out any that are required by eBay, the same page will appear but with the outstanding fields highlighted in red

Private Registration: Enter Information

1 Enter Information   2. Choose User ID & Password   3. Check Your Email

The following must be corrected before continuing:
- First name - Please enter this information.
- Last name - Please enter this information.
- Address - Please enter this information.
- Primary telephone - Your phone number must be at least 10 digits.
- Email address - Please enter this information.

6. Complete the required fields and resend the form

Don't forget

Your User ID is the name by which you will be identified whenever you bid, buy or sell anything on eBay. It is possible to change your User ID once you have created it.

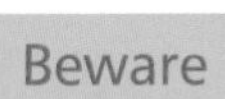

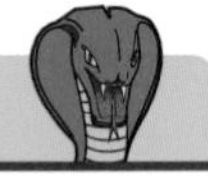

Beware

Do not enter incorrect contact information when registering. If you do, the details will probably be different from those held on your credit card and you will not be able to complete the registration process.

7. If the User ID you have chosen has already been taken you can select one generated by eBay

Choose an eBay User ID

Your User ID identifies you to other eBay members.

- nick9559
- nickv9559
- 9559nickv
- Create your own ID: nickvan2

8. Select one of the suggested User IDs or enter a new one of your own. Click the Continue button

Continue >

9. You need to confirm your identity by entering your credit card details. This is to ensure that the details you have given during registration match those associated with your credit card. Once this has been confirmed, click the Continue button to finalise your registration

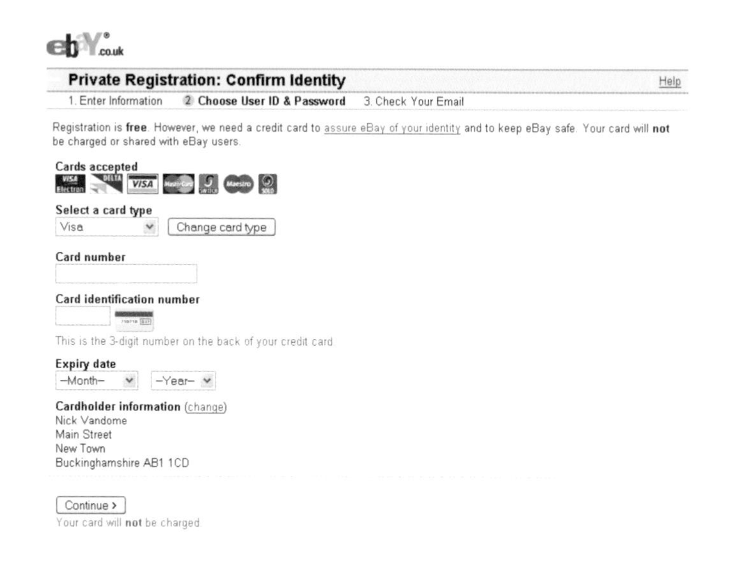
ebay.co.uk

Private Registration: Confirm Identity Help

1. Enter Information 2 **Choose User ID & Password** 3. Check Your Email

Registration is **free**. However, we need a credit card to assure eBay of your identity and to keep eBay safe. Your card will **not** be charged or shared with eBay users.

Cards accepted

Select a card type
Visa | Change card type

Card number

Card identification number

This is the 3-digit number on the back of your credit card

Expiry date
–Month– | –Year–

Cardholder information (change)
Nick Vandome
Main Street
New Town
Buckinghamshire AB1 1CD

Continue >
Your card will **not** be charged

Hot tip

If you forget your User ID, or your password, you can contact eBay from a link on the Sign In page and have it sent to your email address.

Don't forget

eBay has very strict privacy policies to ensure that credit details are only used for verifying your identity. These details will not be divulged to a third party.

Categories

The main objective for most people who visit eBay is to buy or sell items. In both cases it is important to know what you are looking for and where to find it: buyers will want to locate the items they want to purchase and sellers may want to research similar items that are already on sale and see where they have been placed on eBay. One of the ways to look for items on eBay is to browse through the Categories lists. These can be accessed from the home page and then sub-categories can be located. In some instances, several levels have to be navigated until you find the item you are looking for, but it is a useful way to become familiar with the categories of items that are available on eBay. To look for an item using Categories:

Don't forget

Category placement is up to the seller and depends on the selection that they make when they register an item for auction.

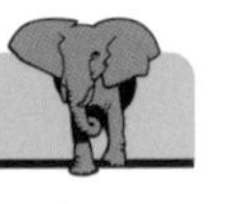

Don't forget

Because of the fluid nature of eBay, the Categories headings do change from time to time, to accommodate changing trends.

1. Click one of the main categories on the home page

Categories
Antiques & Art
Baby
Books, Comics & Magazines
Business, Office & Industrial
Cars, Parts & Vehicles
Clothes, Shoes, Accessories
Coins
Collectables
Computing
Consumer Electronics
Crafts
Dolls & Bears
DVD, Film & TV
Health & Beauty
Home & Garden
Jewellery & Watches

Hot tip

Browsing through the Categories is a good way to get an idea of the types of items available on eBay. However, after a while you may find that a more efficient way of finding items is to use the eBay search facility.

2. Click one of the sub-categories to drill down further through the list

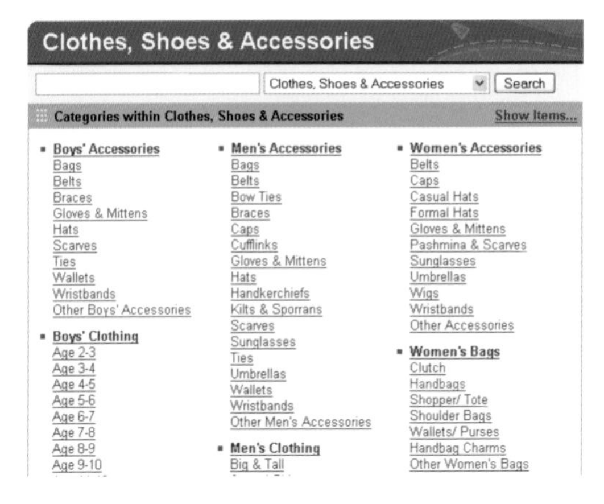

3. A display page shows the current items being auctioned in the selected category

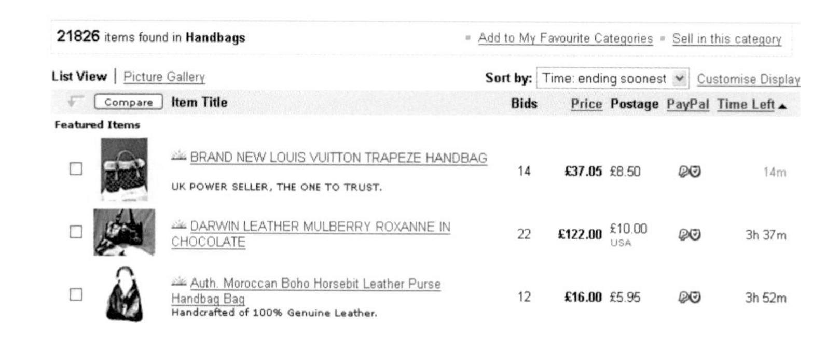

By default, search results pages show the relevant items sorted by the least time left in the current auctions, so that the auctions that are nearest to ending are listed first. This can be changed by clicking on the arrows in the Sort By box and selecting a different option.

4. To refine the search within the selected category, or sub-category, enter a keyword into the Search box at the top of the page

All Items | Auctions | Buy It Now
Gucci | Handbags | Search
Search title and description

5. The search page now displays the results determined by the search keyword criteria

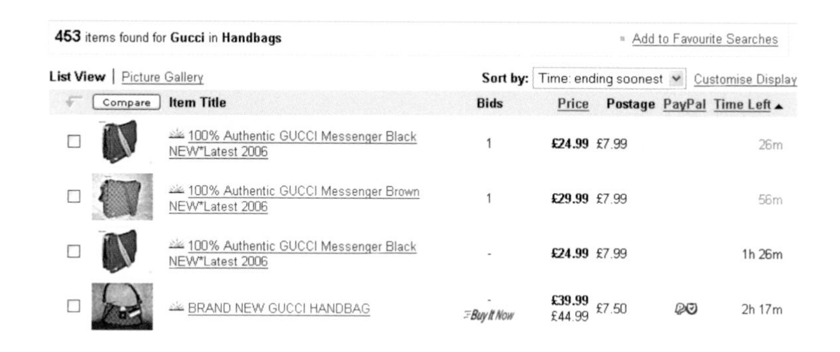

Searching for items

Don't forget

Keyword searches look at the titles and descriptions that have been given to an item by the seller.

Hot tip

If you are having difficulty finding something through the use of keywords, try using alternate spellings, or even misspellings, of the keywords.

Don't forget

Once you have accessed an item from the Search results page you can then participate in the auction by placing a bid. However, you have to be a registered user before you can do this.

Instead of drilling down through the Categories, another way to look for items is to use the eBay search engine. This operates in a similar way to an Internet search engine (such as Google) and, if used accurately, it can be an excellent way of locating specific items. To use the Search facility:

1. The search box is located on the main eBay navigation bar

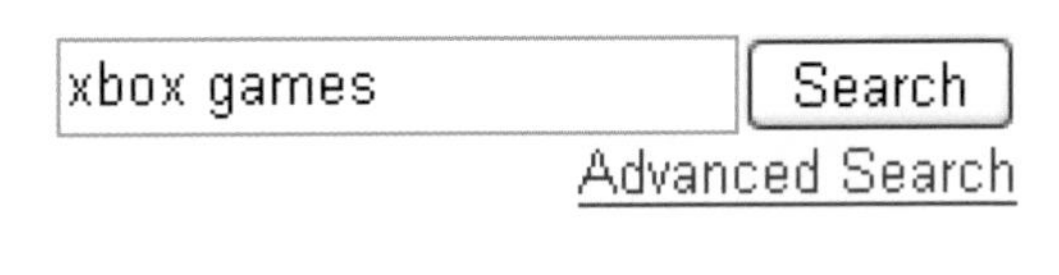

2. Enter a keyword into the box for the item you want to locate, and click the Search button

3. The results page shows the number of matches that have been found and lists them according to the properties that were set in the initial search. It is then possible to view a particular item by clicking on its link

5346 items found for **xbox games** — Add to Favourite Searches

List View | Picture Gallery — **Sort by:** Time: ending soonest — Customise Display

Compare	Item Title	Bids	Price	Postage	PayPal	Time Left ▲
Featured Items						
☐	**Microsoft Xbox 360 Premium Console with games!** Comes with PGR3, Quake 4, Perfect Dark Zero, FIFA 06	12	£455.99	--		1h 08m
Optimize your selling success! Find out how to promote your items						
☐	outrun 2 , x box game	1	£4.00	£2.50		1m
☐	MICROSOFT X-BOX LORD OF THE RINGS GAME (PAL ORIGINAL)	2	£2.21	£1.99		1m
☐	DOOM 3 RESURRECTION OF EVIL. NEW & SEALED XBOX GAME.	-	£11.99	--		2m
☐	ESPN NFL 2K5. NEW & SEALED XBOX GAME.	-	£6.99	--		3m

Refining a search

If too many results are returned from an initial search keyword then you may want to refine the search to limit the results to just the items you want. This can be done by entering additional search criteria:

1. Enter additional words or phrases in the Search box. In this case the minus sign (-) indicates that we do not want any items with the word 'football' in the results

xbox games -football

☐ Search title **and** description

2. The search results reflect the amended search criteria. In this example the number of search results has been reduced from 5346 to 947

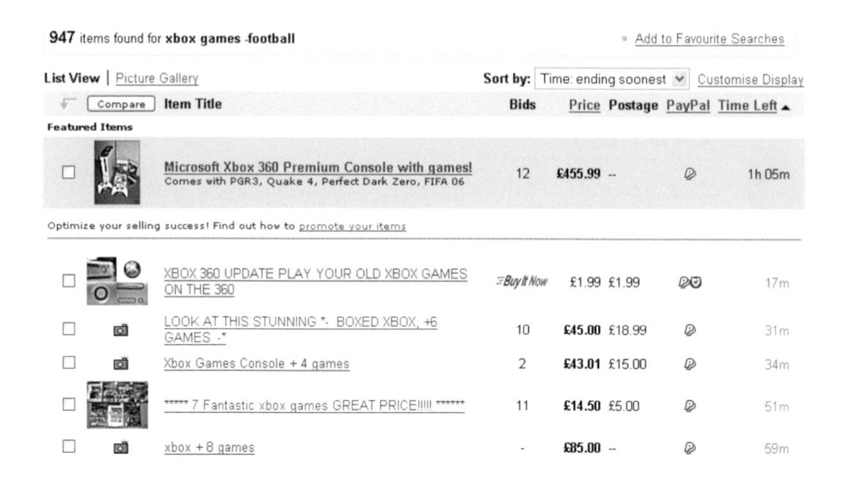

Hot tip

If you want to find an item with an exact phrase, include it within inverted commas, e.g. "Barbie doll outfits". This will ensure a more accurate search. If you want to search for an item with one specific attribute, enter a plus sign before the attribute.

Hot tip

If you want to broaden a search, tick the Search Title and Description box underneath the main search box. This will then search over the description of the item that the seller has provided rather than just the main title. This can result in a significantly greater number of search results.

...cont'd

3. To further refine the search criteria when looking for items, click Advanced Search

Start new search Search
Advanced Search

4. Enter the required details to refine your search. This can include information such as price, location, currency and the condition of goods. Click Search to see the results

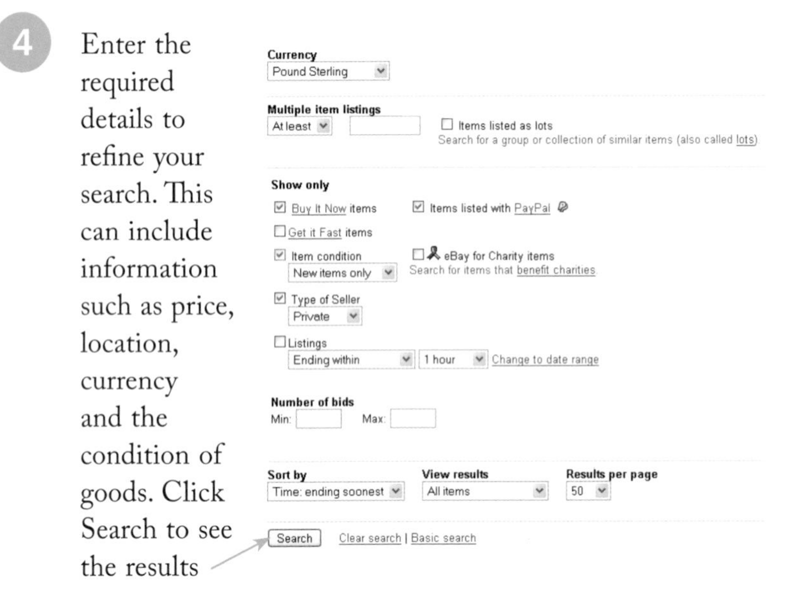

Don't forget

As you become more experienced at searching for items on eBay you may find that you only need to enter the correct keywords and details in the search box. However, the More Search Options box can be useful for specific occasions, such as where you want to limit the results by price.

5. The research results are narrowed down even further

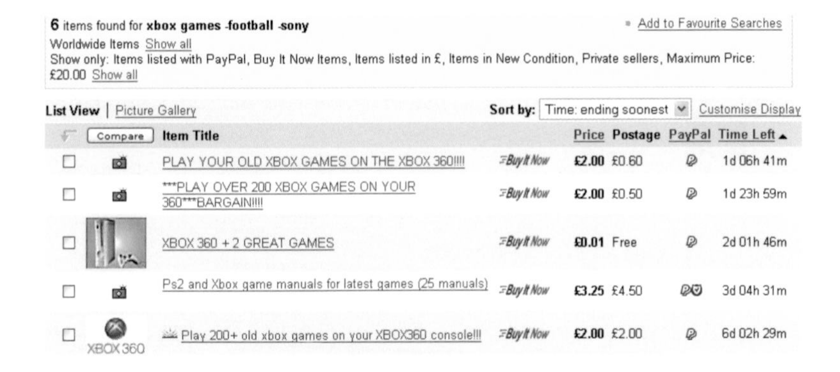

Creating favourite searches

If there are certain items that you want to look for on a regular basis, it is possible to save a search for a particular item and then access it whenever you want, to see how the results for that item have changed over time. To create a Favourite search:

You can create dozens of Favourite Searches; this is a good way of keeping track of what is becoming available for a variety of items.

1. Click the Add to Favourite Searches link at the top of the results page once a search has been completed

Add to Favourite Searches

2. Select the details for how you want the Favourite search to operate and click the Save Search button

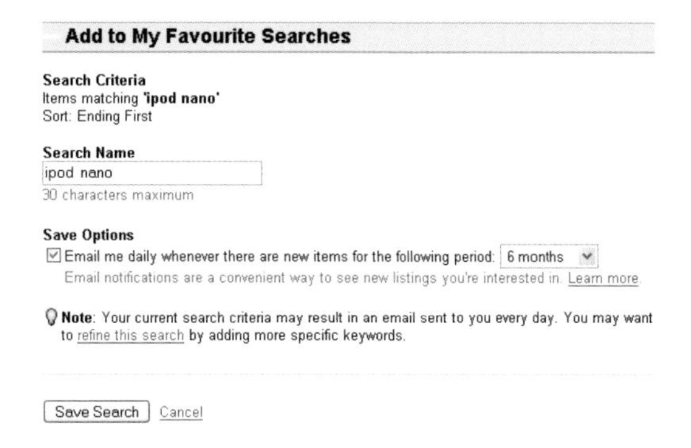

Add to My Favourite Searches
Search Criteria
Items matching 'ipod nano'
Sort: Ending First
Search Name
ipod nano
30 characters maximum
Save Options
Email me daily whenever there are new items for the following period: 6 months
Email notifications are a convenient way to see new listings you're interested in. Learn more
Note: Your current search criteria may result in an email sent to you every day. You may want to refine this search by adding more specific keywords.
Save Search
Cancel

If you tick the 'Email me daily for' box and specify a time frame, you will be sent an email whenever anything new comes up for auction within these search criteria.

3. On the Search page a list of Favourite Searches appears as a drop-down list. Each item can be accessed by clicking the Go button. The results will be displayed for the items at any given time, since they change depending on which auctions are taking place

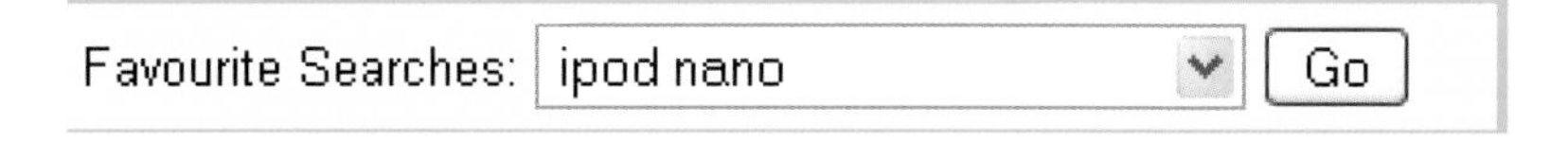

Shops

Due to its remarkable level of success, eBay has become more than just a location for individuals to buy and sell their old record collections or antique teddy bears. There are now a number of speciality sites that are, in effect, online shops. They are run mainly by online retailers rather than individuals, and, although a lot of them conduct auctions for their goods, a large number of items are sold at a set price. The eBay Shops can be accessed from the Speciality Sites box on the home page:

Don't forget

As the popularity of eBay increases, more and more retailers are realising the benefit of having an online presence. This has resulted in a greater number of eBay shops.

1. Click the eBay Shops link to see the range of online shops (this tends to change fairly regularly as online retailers, just like the bricks and mortar ones, come and go)

Speciality Sites
Business Centre
eBay for Charity
eBay Shops
PayPal
Reviews & Guides

Don't forget

Some eBay shops display only Buy It Now items, which means you can only buy them at the price displayed, in the same way as a conventional retailer. Other shops have an auction option as well as Buy It Now. For more information on Buy It Now see Chapter Four.

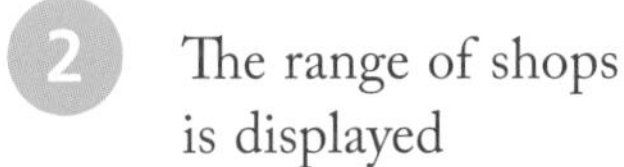

2. The range of shops is displayed

3. Click on a logo to see the goods for that shop

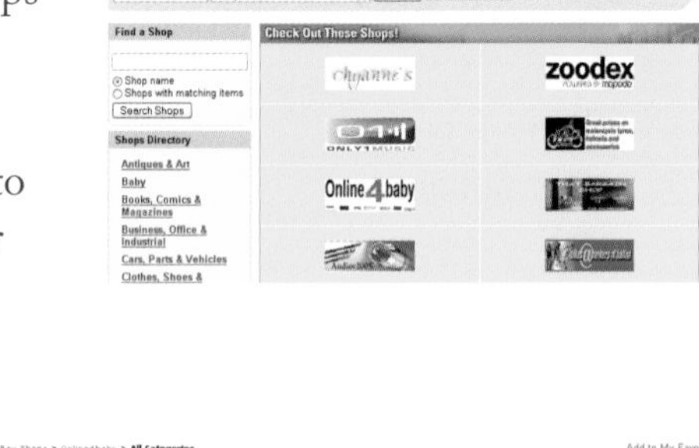

4. Items can be bid on or bought at a set price in the same way as eBay auctions between individuals

Motors

eBay has a special category for shops that are used exclusively to sell cars and related accessories. Since this invariably involves larger sums than a lot of other eBay auctions, a certain amount of caution is required when dealing with this type of transaction. Treat it in the same way as buying a car from a classified advertisement, and if possible go to see the vehicle before you bid for it. Also, have a very careful look at the description of the vehicle and the payment and delivery details. To use the Motors shop, access it from the eBay Motors link on the home page.

1. Enter a vehicle type here to search for a particular model of car, or...

2. Click on a link to see the models for that category

3. The results are displayed in the same way as for any other auction. Click on a link to see details of a particular listed car

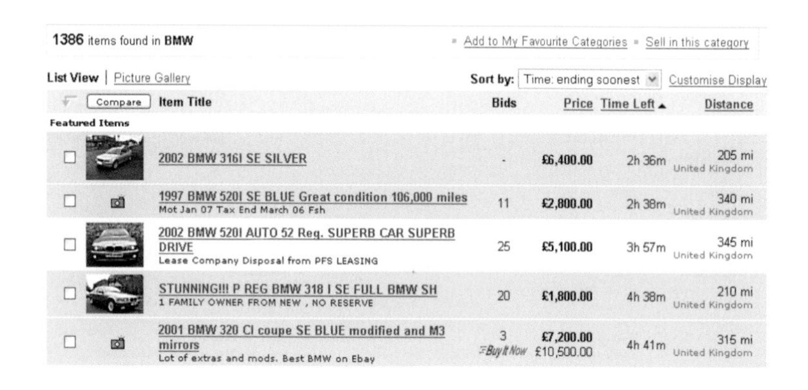

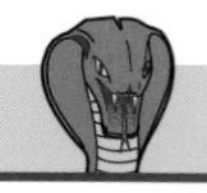

Beware

Because of the large sums of money that can be involved, it is vital that you find out as much as possible about the vehicle in which you are interested and also the person selling it. If you are involved in transactions with large sums of money, there are ways to try and ensure the money is held safely until the deal has been done. This is known as escrow and is looked at in detail in Chapter Ten.

...cont'd

4. The description and auction details are the same as for any other item. Bids can be made by clicking on the Place Bid button

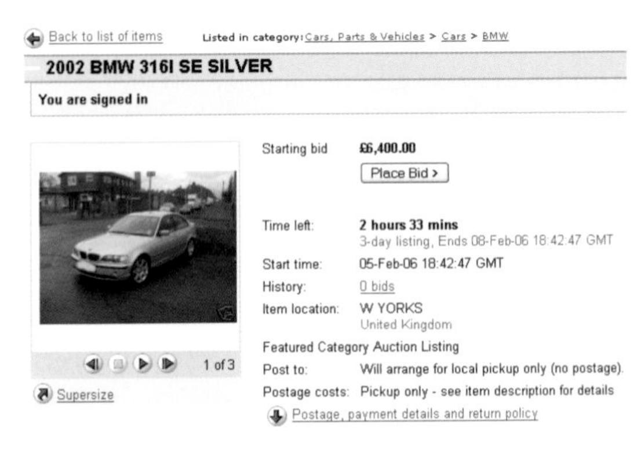

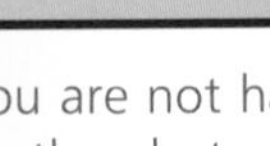

Beware

If you are not happy with the photographs that are displayed for a particular vehicle, or, even worse, there are no pictures, then do not proceed with bidding on the auction.

5. Click under the main picture to view additional pictures of the vehicle

6. Click on a thumbnail image to see larger pictures of the vehicle

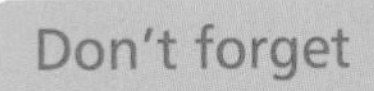

Don't forget

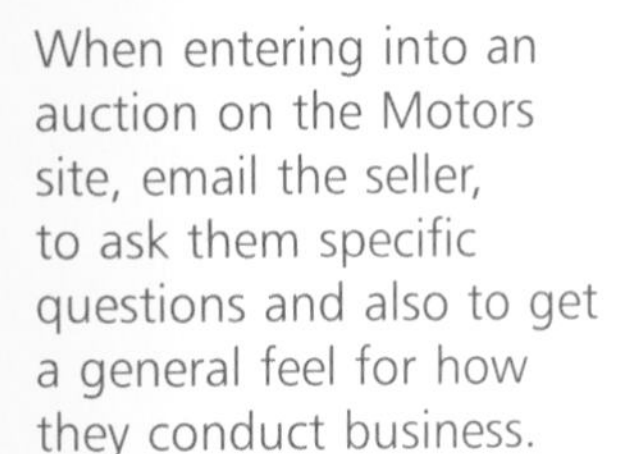

When entering into an auction on the Motors site, email the seller, to ask them specific questions and also to get a general feel for how they conduct business.

7. Scroll down the page to see a full description of the vehicle. It is essential that this is as comprehensive as possible

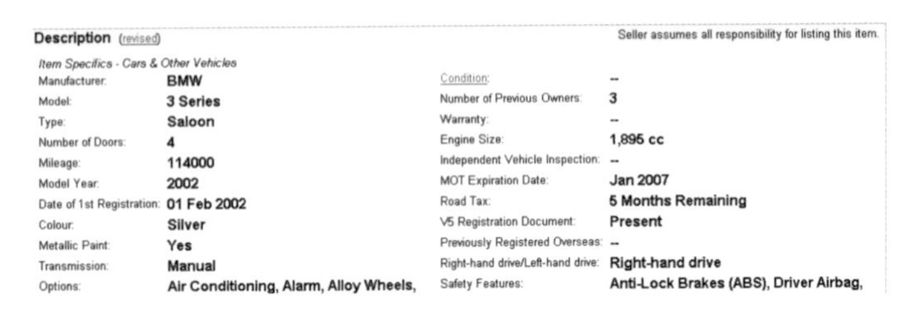

Description (revised) — Seller assumes all responsibility for listing this item.

Item Specifics - Cars & Other Vehicles

Manufacturer:	BMW	Condition:	--
Model:	3 Series	Number of Previous Owners:	3
Type:	Saloon	Warranty:	--
Number of Doors:	4	Engine Size:	1,895 cc
Mileage:	114000	Independent Vehicle Inspection:	--
Model Year:	2002	MOT Expiration Date:	Jan 2007
Date of 1st Registration:	01 Feb 2002	Road Tax:	5 Months Remaining
Colour:	Silver	V5 Registration Document:	Present
Metallic Paint:	Yes	Previously Registered Overseas:	--
Transmission:	Manual	Right-hand drive/Left-hand drive:	Right-hand drive
Options:	Air Conditioning, Alarm, Alloy Wheels,	Safety Features:	Anti-Lock Brakes (ABS), Driver Airbag,

Want It Now

As well as facilities for people selling things, eBay also has an option where you can tell people what you are looking for, in the hope that you can match up with someone who has a similar item to sell. This streamlines the buying and selling process as it removes the need for an auction. This option is known as Want It Now. To use it:

1. Click here under See Also on the home page

See Also

a/A Change text size
Answer Centre
Community Hub
eBay Groups
Gallery
More Tools & Programmes
Seller Tools
Want It Now

2. Click here to post details of an item you want to buy

3. Enter a description of the desired item and click here. The details will then be posted on a Want It Now board which anyone can view. If someone has a similar item to the one in which you are interested they will contact you to conduct a sale

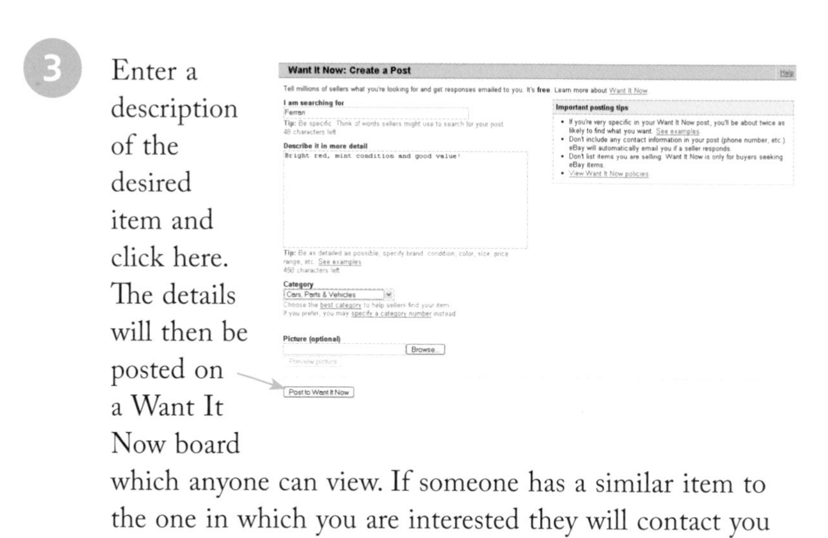

Once items have been placed on the Want It Now page, sellers can view the items or search for specific items. This way, buyers and sellers can match up their requirements.

Charity auctions

Charities are constantly trying to find new ways of generating revenue. eBay has a dedicated site on which charities can promote their causes and sell items to raise funds. These items have often been donated by celebrities and they are frequently signed.

Charities register on eBay to sell items in the same way as anyone else and they have to pay the same fees. However, their auctions receive free promotion and they can create an About Me page to give additional information about the charity and its work.

Don't forget

One of the most questionable categories of item that are sold on eBay is celebrity autographs. There are a lot of these for sale in charity auctions but in general these are more authentic than for standard eBay auctions. Always be careful about autographed items that are not charity auctions.

1. To view charity auctions, click on the Charity Auction link on the home page

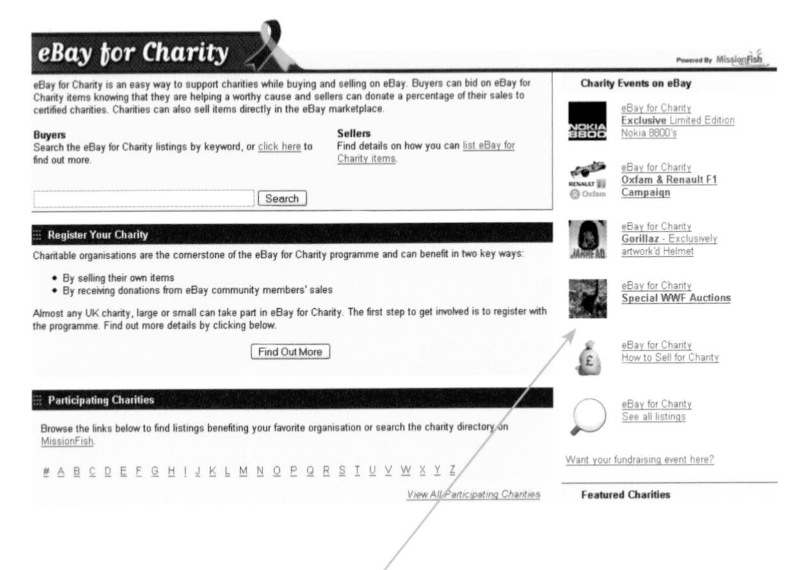

2. Click here to view current charity auctions or select a charity from the alphabetic list

Communities

One of the founding philosophies of eBay is that it should be a community of users who communicated and interacted with one another rather than just processing commercial transactions. Despite its huge financial success, this ethos has remained on eBay and the concept of a huge online community is as strong as ever. In recognition of this, there is an individual site devoted to the eBay community. This includes discussion boards, news and announcements, upcoming events and general community values. To participate in the eBay community:

1. Click the Community button on the main navigation bar

home | pay | site map

Buy	Sell	My eBay	Community	Help

2. Click on one of the links to go to that page within the community. Some areas, such as the discussion boards, have a huge amount of information and comments within them

Connect

Discussion Boards
Discuss any eBay-related topic.

Groups
Share common interests in a public or private format.

Answer Centre
Get quick help from other members.

Don't forget 

The Community area also contains the latest news and announcements from eBay. These cover a variety of topics and reading them is a good way to keep up with the latest developments.

Hot tip 

eBay staff members participate in discussion boards and chatboards and this can be an invaluable source of information regarding the site.

...cont'd

3. Within an area such as the discussion boards it is possible to find out more general information about them, view what is being said by others and also add your own comments

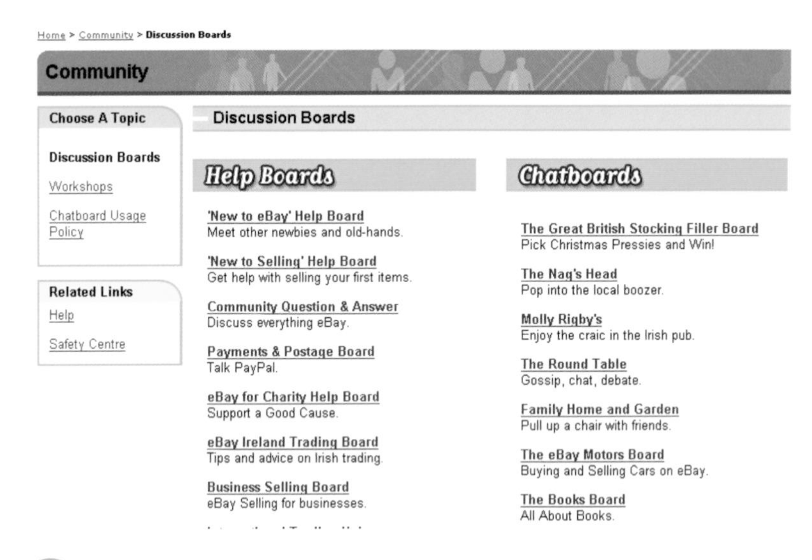

It is perfectly allowable to view the postings on the discussion boards without posting anything yourself. This is known as 'lurking' but it is not generally frowned upon.

4. Click on a link to view a specific discussion board

5. The individual conversation threads are displayed and can be followed by clicking on the links

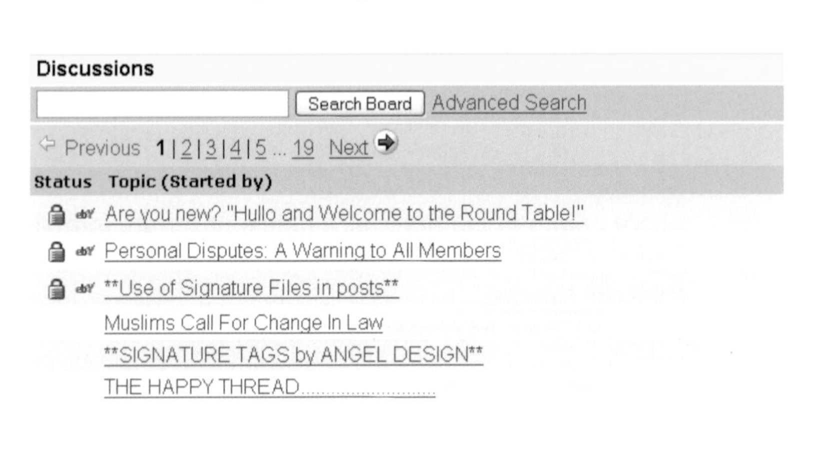

Gallery

When you start selling items on eBay you have to give a title to the item you want to sell, and also a fuller textual description. However, there is also an option for adding a photographic image to show potential buyers what the item actually looks like. This is an important service and one that should always be considered by sellers. One of the reasons for this is that there is an eBay site – the Gallery – that only displays auctions for items where an image has been provided by the seller. This can be accessed from the Gallery link within the See Also box on the home page.

1. Within the Gallery, click on a link on the main page

2. The auctions are displayed with larger images than on regular auction pages

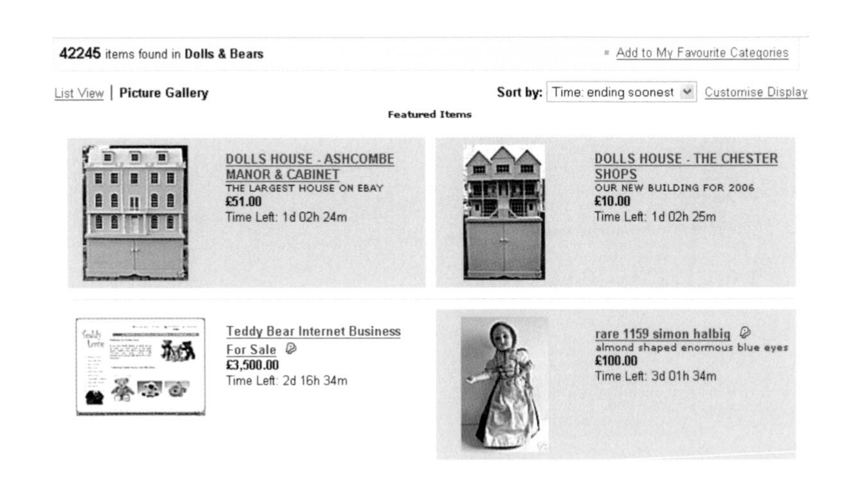

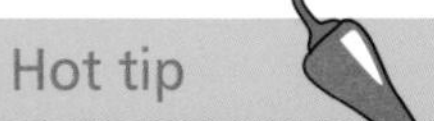

When listing items on eBay, you do not have to include photographic images, but the Gallery shows the importance of doing so. If you do not have an image for your product, then it cannot be included on this page, thus limiting your potential market.

Going, going, gone!

The time limit for eBay auctions varies, but it can be up to 10 days. For some people this is too long to wait and they are only really interested in auctions that are going to end shortly. This means they know they can bid on an item and see almost immediately whether they have won or not. In order to cater for these types of purchasers, eBay offers a service for showing all of the auctions that are about to finish within the next few minutes. These items can be accessed from the home page.

1. Auctions ending soon appear on the home page in a box like this

Hot tip

Refresh the results page to see the time left for each auction change minute by minute.

2. The items listed are current auctions that are about to end. By default, the soonest ending ones are listed first

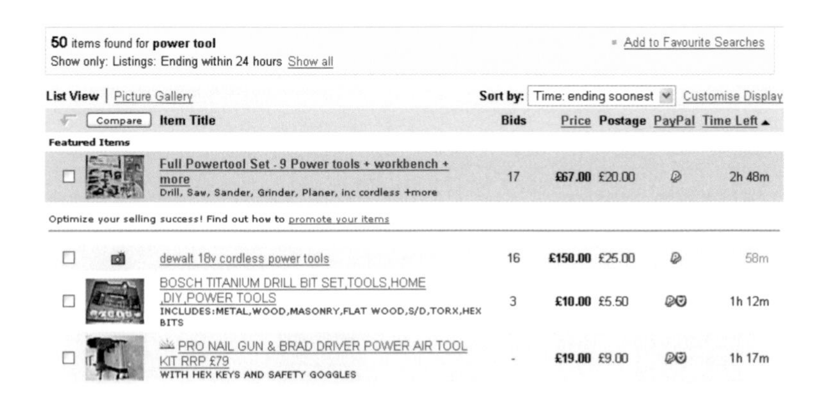

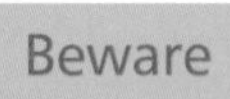

Beware

If you want to bid for an item that has a minute, or less, to go before the end of the auction, you would have to be quick; the process for registering a bid can take a couple of minutes or so, depending on the speed of your Internet connection.

International eBays

eBay is a global phenomenon with sites in over 33 countries around the world. However, in keeping with the community nature of eBay, it is possible to buy and sell items on any of these sites, as long as the seller is prepared to send the items internationally. Once you have registered, or logged in, you can operate on any eBay site in the same way as you would on the one in your own country. To view eBay sites around the world:

1. Select a site from the drop down list in the Global Sites box on the home page

2. Some international sites have a familiar look and feel and they can be used in the same way as any other eBay site

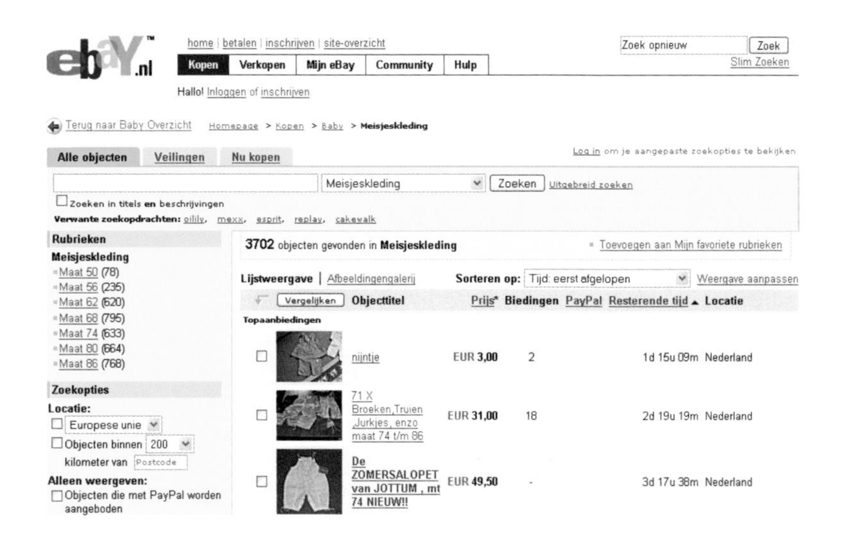

European eBay sites look very similar to the British and US versions. Some sites, such as those in Latin America and Asia, have their own individual characteristics and appearances.

...cont'd

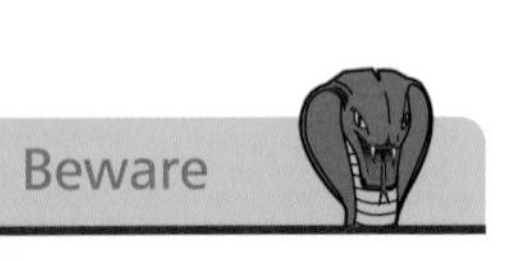

On some international sites, there are translations for the items on sale. However, unless you have a good grasp of the language of the site, think twice before you enter into an auction as the translations are not guaranteed.

3. It is possible to log in to international sites and view areas such as My eBay as you would with your local site. You can also buy and sell items on these sites, but the seller will have to be prepared to ship items internationally

If you are going to bid for items on international sites, make sure that the sellers are prepared to post the items to the country in which you live and not just their own. It is important to do this before you bid rather than afterwards.

4. Some international sites are designed specifically for the region in which they appear, while others are existing auction sites that have been taken over by eBay

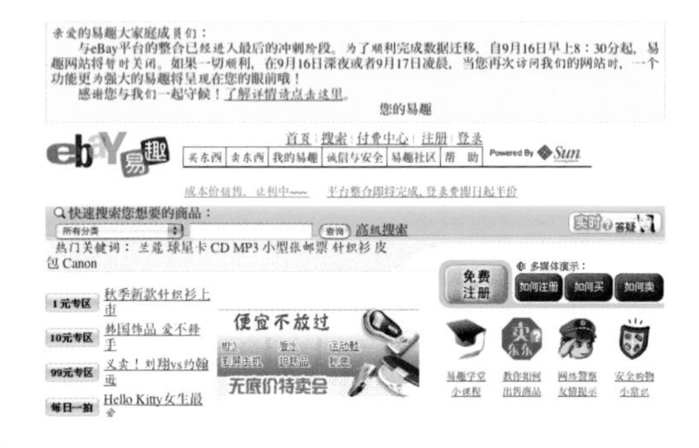

5. When conducting searches, if there are no items in your own locality then there will be a listing for other locations in which similar items are for sale

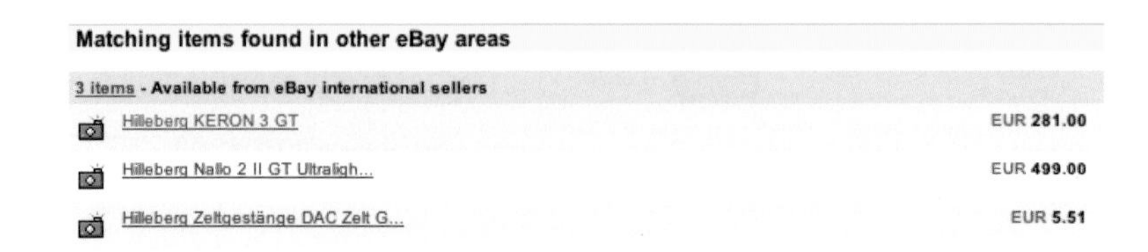

Live auctions

Although eBay auctions can be exciting, particularly towards the end of them, they cannot match live auctions in an auction house. This is addressed by eBay by providing a service where you can view live auctions as they happen around the world. Although this is accessed from the US version of eBay, it can be viewed by users anywhere in the world and, if you sign up, you can bid for the items that are available on the live auction site. To access live auctions:

Don't forget

To bid in Live Auctions, you have to go through a separate login process to the standard one for logging in to eBay.

1. View the eBay site at **www.ebay.com** and click on the Live Auctions link under Categories

2. Click here to view a live auction in progress. Since they happen in real-time, the action is very fast-paced

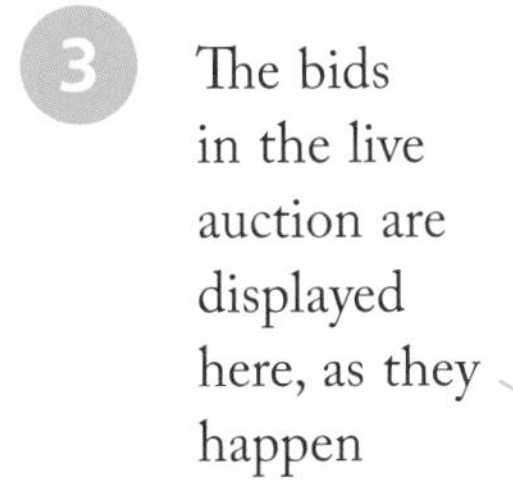

3. The bids in the live auction are displayed here, as they happen

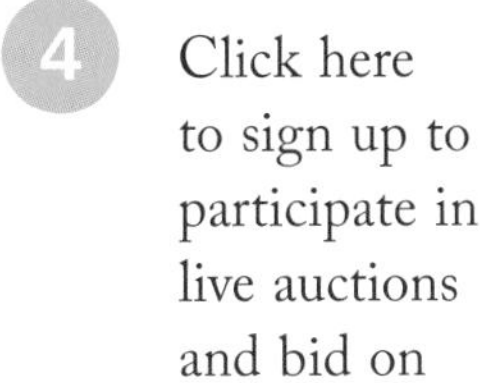

4. Click here to sign up to participate in live auctions and bid on items that appear on the live auction page

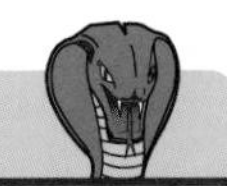

Beware

Live auctions can contain some very high-priced items and they move very quickly. Unless you are very sure that you want to participate, it can be better to view these auctions rather than join in with them.

Finding help

eBay believes in helping users as much as possible and there is a site designed specifically for this. The pages cover all areas of eBay, from buying and selling to what happens when somebody does not pay for items they have won in an auction. To use the help pages:

1. Click the Help button on the main navigation bar

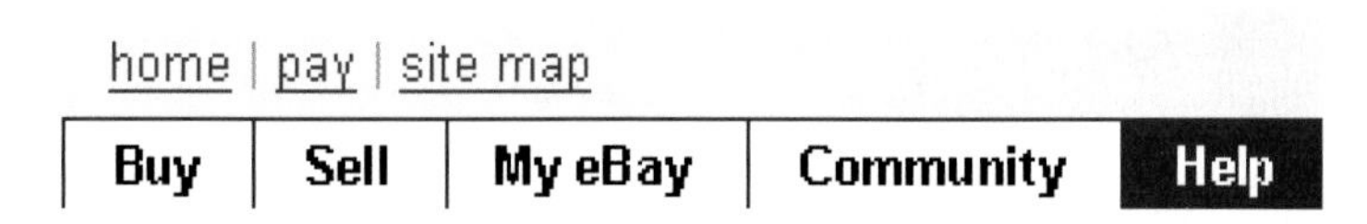

The eBay Safety Centre offers a range of advice about buying and selling safely on eBay. There is a link to the Safety Centre at the bottom of most eBay pages.

2. Select a help topic from the list

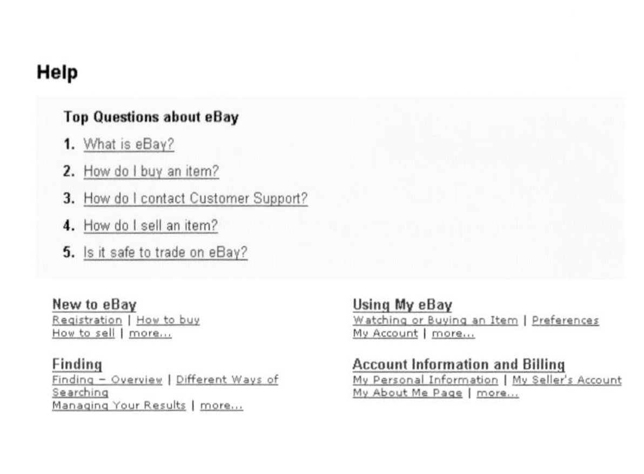

3. Most topics have links to additional information

Managing Buying with My eBay

My eBay 2.0's All Buying page is a complete summary of the status of your buying activities on eBay. You'll see these views on the All Buying page:

- Buying Reminders
- Items I'm Watching
- Items I'm Bidding On
- Items I've Won
- Items I Didn't Win

2 Rules and regulations

The success of eBay is based on the integrity of the site being maintained. Here we look at the various areas covered by the eBay rules, regulations and policies.

General

Given that eBay is a multi-million pound industry it is perhaps surprising that it does not have more problems with fraud and unscrupulous behaviour. However, that is not to say that problems do not occur – they do. But with a bit of groundwork and preparation the careful eBayer can try and minimise the risk of infringing the rules or falling prey to a dishonest buyer or seller.

One of the five founding principles of eBay is that basically people are good. This is supported by the fact that the site has expanded so rapidly and so successfully in such a short period of time. But, as with anything involving money, problems do occur; some of the most common ones are:

- Problems with listings. People, either deliberately or accidentally, give a misleading or fraudulent description.
- Selling dubious items. eBay has an extensive list of prohibited, questionable and infringing items. While a lot of these get filtered out at the time of listing, some of them get through the net. If you are in any way unsure about whether an item should be on eBay or not, do not attempt to buy or sell it.
- Sellers not sending the item as it is described. Sometimes sellers advertise items that are nothing like the product which the winning bidder receives. In some case this is due to laziness or simple inaccuracy, but in others it can be because of fraud. Due to the robust feedback system, fraudulent sellers usually do not last too long, but this is of little comfort if you have already fallen prey to one of them.
- Buyers not sending any payment. There are a number of reasons for this, but the golden rule is that you should never send an item until you have received payment for it. If the payment is by cheque, wait until it has cleared before you send the item. While non-payment can be annoying, it is still possible to re-list an item which has not been paid for.

Don't forget

As far as the legality of transactions on eBay is concerned, individuals are subject to the law of the country of the site on which they are conducting business.

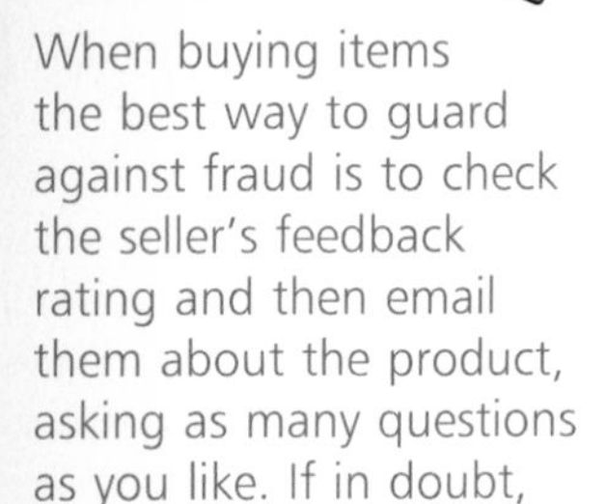

Hot tip

When buying items the best way to guard against fraud is to check the seller's feedback rating and then email them about the product, asking as many questions as you like. If in doubt, do not proceed.

Prohibited items

When you look at the list of categories of items for sale on eBay you may think that it is implausible for any items to be prohibited, but this is far from the case. eBay take matters of the law very seriously and there is an extensive list of items that are banned; this list is non-negotiable, and if you try to list one of these items then you will be subject to eBay's own sanctions and you may even face prosecution. Some of the prohibited items include credit cards, firearms, lock-picking devices, human parts and remains, lottery tickets, prescription drugs and any stolen items. A full list of prohibited items can be viewed on the eBay Rules and Policies page, which can be accessed from the link at the bottom of most eBay pages. For each prohibited item there is a link that contains an explanation of why that item is prohibited and, in some cases, exceptions to the general rule.

Don't forget

There is a large number of eBay staff who check listings, but some prohibited items do occasionally slip through.

Fireworks

Fireworks or pyrotechnic devices may not be listed on eBay.

Pyrotechnic devices include "any combination of materials, including pyrotechnic compositions, which, by the agency of fire, produce an audible, visual, mechanical or thermal effect designed and intended to be useful for industrial, agricultural, personal safety, sporting, or educational purposes."

Examples of fireworks and pyrotechnic devices that may not be listed on eBay include, but are not limited to: blank ammunition, fireworks and fireworks kits, aerial bombs, booby traps (pull string), bottle rockets, chasers, dayglo bombs, firecrackers, fountains, nitro poppers, party poppers, roman candles, skyrockets, smoke balls, smoke bombs, snap caps, snappers, sparklers, sparks, and torpedoes.

Community Watch
eBay Community Watch - Community members can report prohibited, questionable and infringing items to eBay.

Related Help topics

- Prohibited and Restricted Items Overview
- Firearms

Don't forget

If you enter a search keyword for a prohibited item, you may find a number of results. These will probably be related items, such as a book about a prohibited item, rather than the prohibited item itself.

Since each eBay site is subject to the law of its host country, there may be slight differences between their lists of prohibited items. For instance, the list for the US may be different to that for the UK due to legal and political considerations. Before you start selling on eBay, have a good look at the list of prohibited items: if you do transgress them then ignorance is not an excuse. Similarly, when purchasing items, if you think they may be prohibited, contact eBay to tell them about it and ask them for advice about how you should proceed. (See page 40 for details of how to report infringements on the site.)

Questionable items

As well as prohibited items, there are equally extensive lists for questionable and infringing items, both of which should be read in full before you start selling.

Questionable items are regarded as those that can be sold only under certain circumstances. In some cases, each item of this type is judged on its merits. One of the questionable items is Adult Material, which is a good example of how the interpretation is not always clear-cut. The default position is that eBay does not allow the listing of erotica or sexually orientated material. However, if this type of material is included as part of another item then it may be allowed, as long as it is a small part and the main item does not fall into this category. Similarly, some artwork could be listed depending on the context.

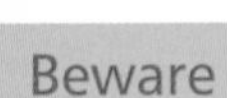
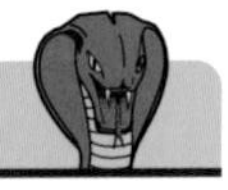

Beware

Some software programs appear for sale with a description stating that they should only be used as backups if you have bought the original program. This is definitely a grey area and buyers should treat it with caution.

Infringing items are those that could infringe an existing copyright or trademark. The most common examples of this occur with software programs, video games and films. For instance, it is permissible to list a genuine, unused version of a software program, but unauthorised copies cannot be listed. The same applies for pirated video games and films.

As with prohibited items, questionable and infringing items are subject to the law of the country in which the site is located. This can create some interesting regional issues. For example, on **eBay.co.uk** consider the sale of British Titles (which comes under the Questionable category). It is not permissible to sell or buy titles such as Duke and Duchess, Earl and Countess or Viscount and Viscountess, but it is allowable to list Scottish Feudal Baronies as long as it is stated that they are for Scotland and they are clearly and accurately described.

Because of the size of the site it is impossible for eBay to police every single auction. However, if you do get caught repeatedly breaking the rules you will probably be given three warnings and after that you will have your account suspended. This means that you will not be able to trade (either buying or selling) until the suspension is lifted.

Listing policies

One of the ways in which eBay tries to enforce its policy on prohibited, questionable and infringing items is through its listing policies. These form an extensive list; some of the areas covered are:

- How to make sure that you list your product so that it appears in the correct category.
- How to list items if you are selling duplicate items within the same auction.
- Creating listings where you are offering a prize or free gift for the person who wins the final auction.
- The types of listing that are not permitted, such as those promoting prize competitions or lotteries, those which try to avoid paying the standard eBay fees, and those which include language that is abusive in terms of being racist, hateful, sexual or obscene.
- The eBay links policy. Details the types of links that can be used within a listing (usually to other areas within eBay) and those which cannot be used.

Full details of the listing policies can be found by following the Listing Policies link on the Rules and Policies page:

Don't forget

Selling duplicate items within the same auction is known as a Dutch auction and this is looked at in more detail in Chapter Four.

1. Click on a link to see details of a particular item within the Listing Policies

Listing Policies for Sellers

Familiarise yourself with eBay's listing policies before posting an item. You'll have a safe and fun selling experience knowing that your listing complies with community standards and guidelines.

Listings that violate eBay's policies may result in disciplinary action. This action may include a formal warning, the ending of all violating listings, or even temporary or indefinite suspension of a user's account. eBay will consider the circumstances of an alleged offence and the user's trading records before taking action. In most cases, eBay will credit all associated fees when a listing is ended.

Listing Policies:

- About Me Guidelines
- Bonuses and Prizes
- Catalog Sales
- Categorisation
- Choice Listings
- Circumventing Fees
- Duplicate Listings
- Excessive Postage/Packaging
- Giveaways, Raffles, and Prizes

User Agreement

Most people who have used computers will have come across User Agreements before, either when installing software or when purchasing items from an online retailer. In many cases, people simply click the Accept button when faced with these types of agreements and then move on to something more interesting. However, in the case of eBay and its User Agreement it is a good idea to read it in full. This is so that you understand what is expected of you while you are on the site and how you can expect others to behave towards you too. Also, the User Agreement contains a lot of useful general information about eBay, and it is a good introduction to the Do's and Don'ts of the site. To access the User Agreement:

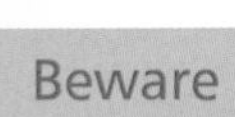

At the bottom of the User Agreement is a statement that says use of the site constitutes acceptance of the User Agreement. It is not something for which you actively sign up.

1. Access the Rules and Policies page and click on the User Agreement link
2. If you have not already done so, take the time to read the User Agreement in full

Your User Agreement

Welcome to the User Agreement for eBay.co.uk ("User Agreement"). The services available at http://www.ebay.co.uk are provided by eBay International A.G. ("eBay", "we", "us" or "our"), located at Helvetiastrasse 15/17, 3005 Bern, Switzerland.

This User Agreement describes the terms and conditions applicable to your use of our services available under the domain and sub-domains at http://www.ebay.co.uk(the "Site"). If you do not agree to be bound by this User Agreement, you may not use or access our services.

You must read, agree with and accept all of the terms and conditions contained in this User Agreement, which includes those terms and conditions expressly set out below and those incorporated by reference, before you may become a user of the Site. We strongly recommend that, as you read this User Agreement, you also access and read the information contained in the other pages and websites referred to in this document, as they contain further terms and conditions which apply to you as an eBay user. Please note: underlined words and phrases are click-through links to these pages and websites.

Privacy

Another issue which is taken very seriously on eBay is privacy. Due to the amount of personal information that a member has to give before they can register on eBay (name, address, telephone number, email address and credit card details), it is imperative that this information is kept strictly for those people who require it and no-one else.

The privacy policy contains details of the type of information collected by eBay, what is done with it, who it is disclosed to and how you can change your personal details. To review the Privacy Policy:

1. Access the Rules and Policies page and click on the Privacy Policy link

2. Click on a link to view that item in the policy, or scroll down to read the whole policy

Contents

An Important Note About Children
Information We Collect
Our Use of Your Information
Our Disclosure of Your Information
Advertisers
eBay Community
External Service Providers
Internal Service Providers
Other Corporate Entities
Legal Requests
Control of Your Password
Accessing, Reviewing and Changing Your Personally Identifiable Information
Other Information Collectors
Security
Notice

One important issue that is stressed in the Privacy Policy is that eBay cannot be used by children under the age of 18. However, children can use the site if they are supervised by an adult, but they are not allowed to submit any personal information.

Reporting infringements

Due to the number of items for sale on the site, it is impossible for eBay to police every single auction to make sure that none of the rules are being broken. In an effort to preserve the integrity of the site as much as possible, eBay relies to some extent on the community of users to keep an eye out for prohibited, questionable or infringing items. It is in everyone's best interest to monitor the site and maintain its reputation for good quality, legal and genuine products.

One avenue for reporting suspect auctions is through the Community Watch Contact form, which can be found by following the Report a Problem link on the Safety Centre page. This form can be used to report a number of issues, including prohibited items. To use the form:

eBay takes a proactive approach to infringements of the rules and they check through as many auctions as possible. If prohibited, questionable or infringing items are listed for sale there is a reasonable chance that it will be spotted and the auction will be stopped. Disciplinary action may then be taken by eBay against the seller.

1. Select a category on which you want to report, and select the relevant sub-categories

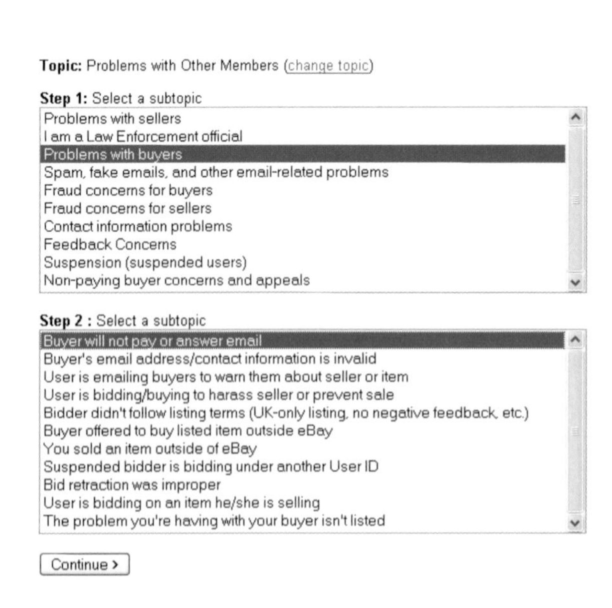

Topic: Problems with Other Members (change topic)

Step 1: Select a subtopic

Problems with sellers
I am a Law Enforcement official
Problems with buyers
Spam, fake emails, and other email-related problems
Fraud concerns for buyers
Fraud concerns for sellers
Contact information problems
Feedback Concerns
Suspension (suspended users)
Non-paying buyer concerns and appeals

Step 2 : Select a subtopic

Buyer will not pay or answer email
Buyer's email address/contact information is invalid
User is emailing buyers to warn them about seller or item
User is bidding/buying to harass seller or prevent sale
Bidder didn't follow listing terms (UK-only listing, no negative feedback, etc.)
Buyer offered to buy listed item outside eBay
You sold an item outside of eBay
Suspended bidder is bidding under another User ID
Bid retraction was improper
User is bidding on an item he/she is selling
The problem you're having with your buyer isn't listed

Continue >

2. Click Continue to send your report

3 Getting started

eBay is known for its auctions, which form the bedrock of the site. This chapter looks at the basics of auctions on eBay and shows the type of information that is displayed. It also covers the My eBay section, which is an individual page that can be tailored to your own requirements.

Icons

eBay makes extensive use of icons to provide visual shortcuts that indicate the status of various elements on the site.

Don't forget

You may have to log in again if you are inactive for a period of time during a session. This is a security measure to try to ensure that no one else can use your account if you are away from your computer.

Member icon

The first icon that you will probably encounter is the new member icon. This shows other members that you are new to the eBay community and are still finding your way. It also shows that you could be a little naive when it comes to some of the finer points of buying and selling on eBay. For most members this is not a problem, since everyone has to start somewhere, but for some it may be an issue. The new member icon stays next to your name for 30 days and then it disappears. The new member icon looks like this:

Don't forget

Users who have recently registered on eBay are known within the community as 'newbies'. While this can be used as a condescending term by some members, it is not generally regarded as derogatory.

Another icon you may come across is the Change of User ID icon. This means that the member has changed their original User ID. In most cases this is perfectly legitimate, but it can be used as a way to try and hide a previously bad trading reputation. The Change of User ID icon looks like this:

Feedback icons

Feedback is vital for assessing a seller's, or a buyer's, reputation and there are numerous icons that denote a member's overall feedback rating (that is, the combination of positive and negative feedback). This is done with a star rating system. Once you have an overall positive rating of 10 or over you will receive a yellow star. This progresses up to a selection of shooting stars, the highest being a red shooting star, which indicates an overall rating of 100,000 or over.

Don't forget

Power Sellers have to maintain an overall feedback rating of 98% or higher. There are different levels, based on either gross volume of takings or quantity of items sold over a three month period. This ranges from £750 or 100 items for the Bronze level, to £95,000 or 500 items for the Titanium level.

Seller icons

Some sellers make selling on eBay a full-time occupation. One way to assess the status of a seller is whether they have a Power Seller icon next to their name. The Power Seller icon indicates a high volume of sales each month, and looks like this:

Power Seller

Accessing an auction

Understandably, given the nature of the site, the first thing that most new eBay members want to do is to have a look at an auction. Before jumping in and starting to bid on items, it is a good idea to have a sound understanding of the auction page and how it operates. The first step is to search for items as shown in Chapter One; you will then be able to view the current auctions for matching items:

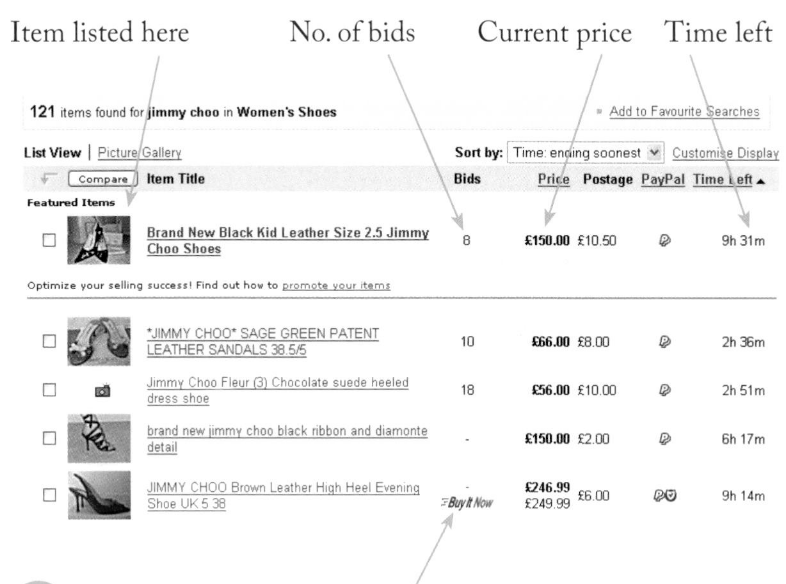

1. Click on Buy It Now to purchase an item without having to bid for it. This enables you to buy the item quickly without entering into the auction. However, make sure it is at a price with which you are happy and not more than you would have been prepared to go up to in an auction

2. To view all of the items with images, click on the Picture Gallery link

List View | Picture Gallery

Don't forget

Items that appear at the top of a search results page are known as Featured Items. These are displayed more prominently than subsequent items, but sellers must pay an additional fee for this service. For more information about Featured Items, see Chapter Seven.

Don't forget

Although the Picture Gallery lists all of the items with images, text-only items are also included, but at the very bottom of the list.

The auction page

Once you have found a specific item in which you are interested, you can access it by clicking on the link next to it on the search results page or on the image (if it has one). This takes you into the auction page for that item. This displays a variety of information about the current auction and enables you to place a bid.

Don't forget

Some auctions have a reserve price attached to them. This is a minimum price that the seller will accept for the item. If the minimum price is not met at the end of the auction then the item will not be sold. The reserve price is not displayed to bidders.

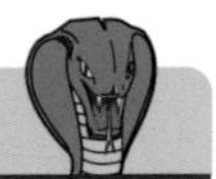

Beware

Don't rush in and bid on the first item you see. Check other similar items and see how much time there is left for a particular auction. It is not a good idea to let other bidders know that you are very keen on a particular item.

No. of bids so far   Time left   Current high bid

Brand New Black Kid Leather Size 2.5 Jimmy Choo Shoes
Bidder or seller of this item? Sign in for your status
Current bid: £150.00 (Reserve met)
Place Bid >
Time left: 9 hours 22 mins
7-day listing, Ends 09-Feb-06 20:17:39 GMT
Start time: 02-Feb-06 20:17:39 GMT
History: 8 bids (£20.00 starting bid)
High bidder:
Item location: Otford, Kent
United Kingdom
Featured Category Auction Listing
Post to: United Kingdom
Postage costs: £10.50 -- Royal Mail Special Delivery
Postage, payment details and return policy
Supersize

1. Scroll down the page to see a full description of the item and any images that have been posted

RELUCTANT SALE!
These gorgeous black kid leather Jimmy Choo shoes called "LIDO" are in fantastic condition. They have been worn indoors ONLY once and have been kept in the original box with tissue paper and the original Jimmy Choo dust bag. Original price £340.00 (receipt still available!)
Larger Picture
Select a picture

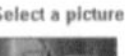

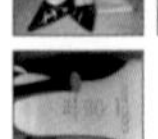

2. Scroll down the page to view postage details. You should read these carefully to make sure the seller will post the items to where you live, and also to check that the postage is not prohibitively expensive

Postage, payment details and return policy

Postage Cost	Services Available	Delivery Time*
£10.50	Royal Mail Special Delivery	1 working day

*Sellers are not responsible for delivery time. Delivery times are prov
and may vary with package origin and destination, particularly durinc

Will post to United Kingdom.

Postal insurance
Not offered

3. Scroll down the page to check the payment details. These will list the ways in which the seller will accept payment for the item

Payment methods accepted

- 

- Personal cheque
- Postal Order or Banker's Draft

Learn about payment methods.

4. If you are happy with all of the details on the auction page and you are still interested in bidding for the item, click on the Place Bid button. This will take you to the next stage of the bidding process and enable you to submit your own bid

Place Bid >

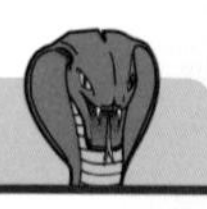

Beware

It is particularly important to check the postage for cheaper items. There is no point in getting a bargain and then having to pay an exorbitant fee in postage. Some sellers have been known to do this deliberately, to make their item appear more attractively priced and also to avoid paying a higher level of fees, a practice that is prohibited by eBay.

Don't forget

Bidding is looked at in greater detail in Chapters Four and Five.

Seller information

At the top-right of the main auction page there is a box containing information about the seller. This is very important as it contains feedback about their previous trading history and it also enables you to contact them directly with any questions you may have about the auction. To access information about a seller:

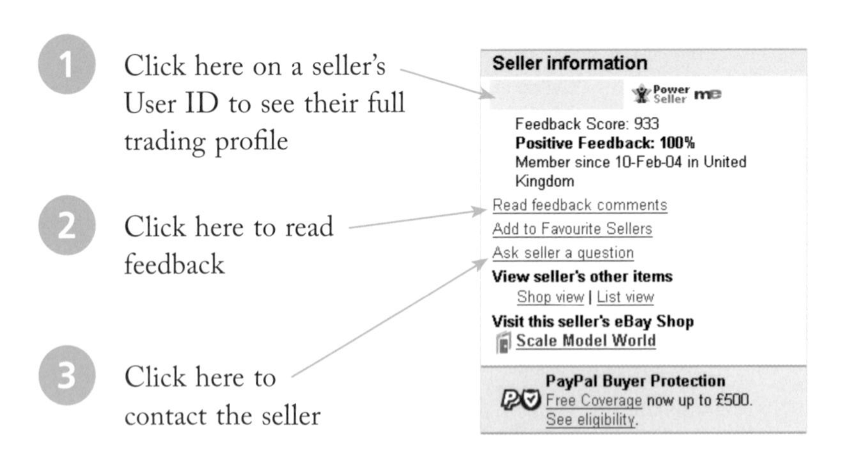

1. Click here on a seller's User ID to see their full trading profile
2. Click here to read feedback
3. Click here to contact the seller

It is always worthwhile to read a seller's feedback. This not only gives you a good sense of their trading history, it is also a good way to familiarise yourself with the type of feedback that is left on the site.

4. The seller's full trading profile contains a summary of the feedback that they have received and also a list of the individual comments

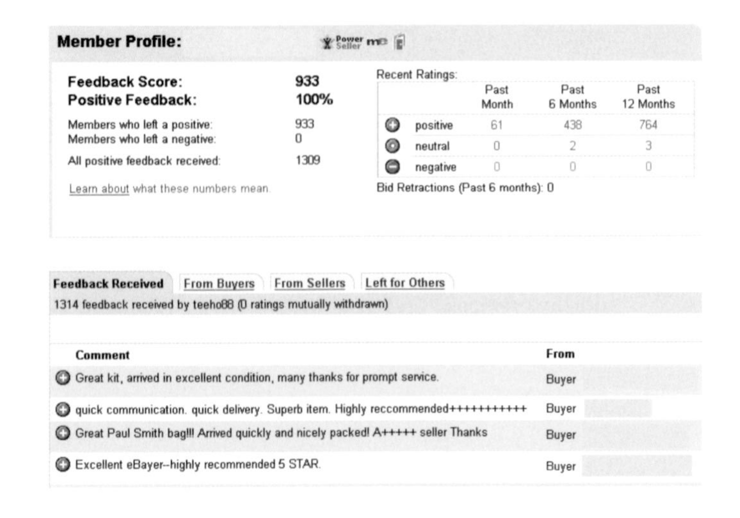

My eBay overview

Even if you only have an initial passing interest in eBay, you may be surprised at how quickly you fall under its spell. Before long you may be actively involved in several auctions, keeping a watchful eye on a few more, conducting your own auctions and trying to keep track of all of the financial comings and goings. Trying to keep up with all of this on different pages could become a logistical nightmare, so eBay has introduced a feature to allow members to keep track of all of their online transactions and activities. It is called My eBay and it is like a diary for everything you are involved with on the site. A My eBay page is an invaluable guide for your life on eBay. To begin creating your My eBay page:

1. Click the My eBay button on the main navigation bar

2. A summary of your eBay activities is listed here as well as links to various options within My eBay

3. Click here to view the latest news and announcements from eBay. These are automatically included on My eBay pages when they are released. Each link takes you to the relevant news item or announcement. Since eBay is such an active and evolving site, this is a good way of keeping up with what is happening

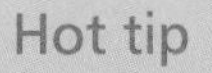

Hot tip

Set My eBay as a favourite on your browser, or even as your home page. You will still have to log in to access it, though.

My eBay page

As you start becoming more involved with eBay, your My eBay page will contain more items corresponding to your activity on the site. These include details of items you are watching, auctions in which you are involved, items you have won or lost in auctions, and a financial history of your dealings on eBay.

To change the time period for the Items I've Won section, click on the Period box and select a different time period.

1. When you have won items in an auction, the details will be displayed within the Summary of your My eBay page:

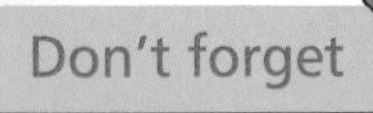

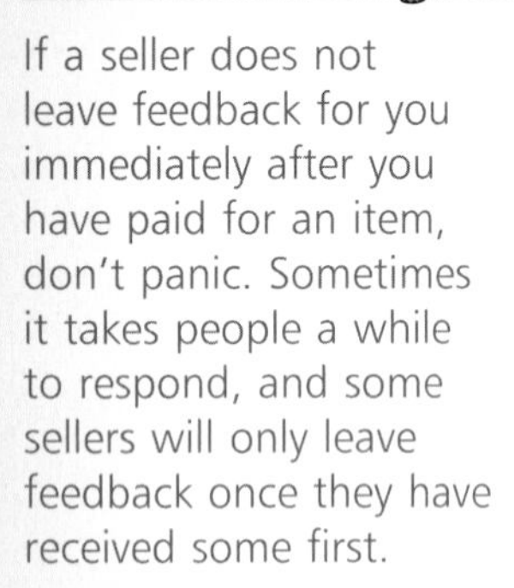

If a seller does not leave feedback for you immediately after you have paid for an item, don't panic. Sometimes it takes people a while to respond, and some sellers will only leave feedback once they have received some first.

2. This icon denotes that an item has been paid for

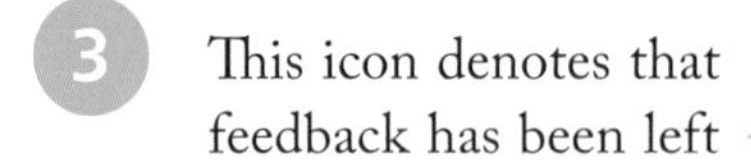

3. This icon denotes that feedback has been left

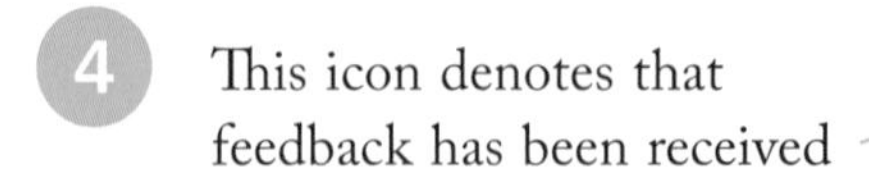

4. This icon denotes that feedback has been received

5. Information on how much you have spent on auctions is listed below the auction details

Watching items

Before you start participating in auctions, it is a good idea to have a look at a few that are selling the types of products in which you are interested. This will give you a good idea of the type of condition they are in, the price ranges in which they are sold and the number of bidders there are for certain items. This last point is important because if there are a lot of people interested in a certain type of product then this could result in some keenly fought auctions and, more importantly, some high prices. As part of the research into how certain items are performing it is a good idea to watch a few auctions to see how they progress. Rather than having to do this individually with a lot of different auctions, it is possible to watch multiple auctions from within My eBay. To do this:

Don't forget

Up to 30 items can be listed to be watched at any one time from within the My eBay page.

Hot tip

The details of watched items update whenever the My eBay summary page is refreshed in the browser window.

1. Open an auction in which you are interested

2. Click the 'Watch this item' link at the top of the auction page

Watch this item in My eBay

3. The selected auction is displayed on your My eBay page

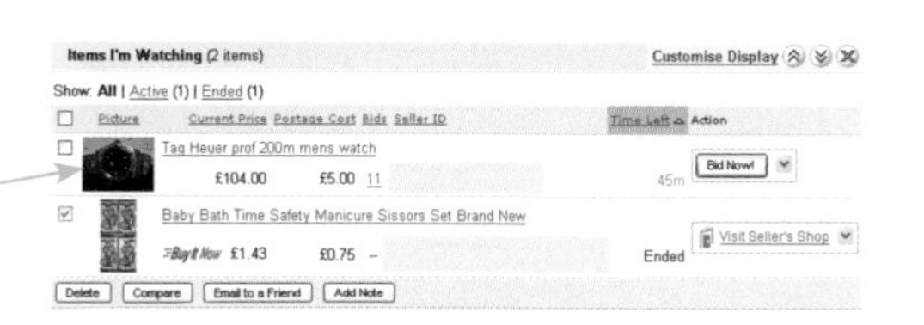

Comparing auctions

A technique that can be even more useful than watching individual auctions is to compare similar ones side by side. This can be done within the My eBay summary, but a more comprehensive option is to view auctions using the Compare function. To do this:

1. Add two or more items to be watched in My eBay

2. Click here to compare the auctions in more detail

3. View the details of similar items next to each other. This is an invaluable option for comparing areas such as bid activity, price, postage costs and payment methods

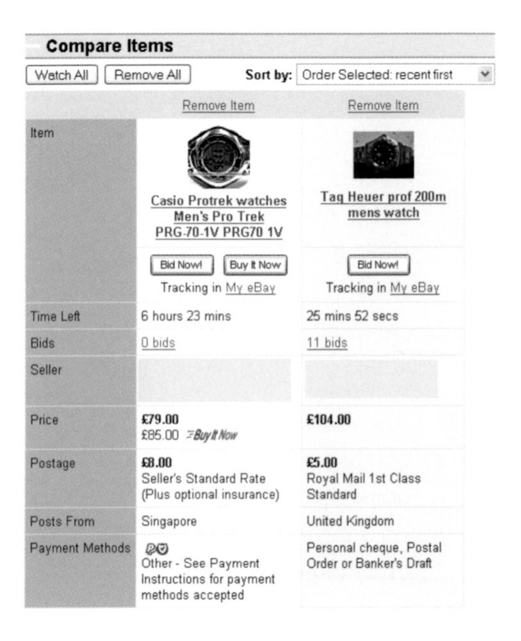

Hot tip

If you are watching multiple items and you only want to compare some of them, check on the boxes next to them to ensure that these are the only ones included.

Hot tip

Another way to compare auctions is to open them in individual browser windows. However, the Compare method is more efficient for a general overview.

Adding notes

Since you may be watching a variety of items for different reasons, it is useful to be able to leave notes next to specific items. These can be used for reminders, for example a time or date at which to bid, or just for general information, such as a suggestion to look up similar items from the same seller when the current auction finishes. To add notes to items that are being watched:

1. Check on the box next to a watched item

2. Click here to add a note

3. Add a note about the item and click Save

4. The note is added here on the watched item within My eBay

Don't forget

Notes that are left next to items you are watching are visible only to you; no other users can see them.

Viewing won items

Whenever you have completed a successful auction and won an item, the details will be displayed in the My eBay summary. The details displayed here can be customised to increase or decrease the number of columns of information on display:

To delete an item from the Items I've Won display, tick the box next to the Seller's ID and click Remove.

1. Details of items you have won are displayed within the My eBay summary. Click here to customise the display

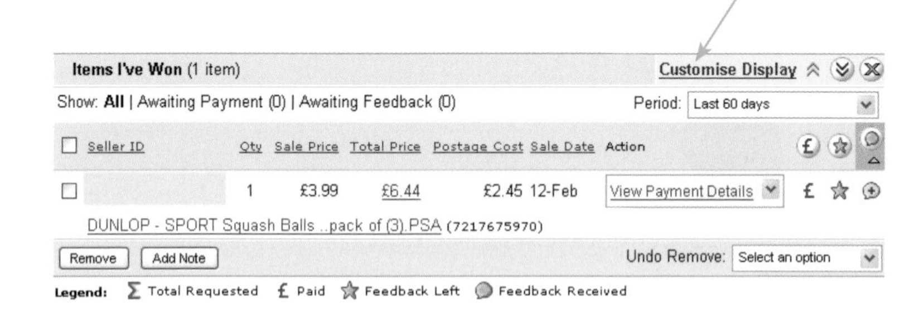

2. Select an item to be included under Item I've Won and click here for it to be displayed

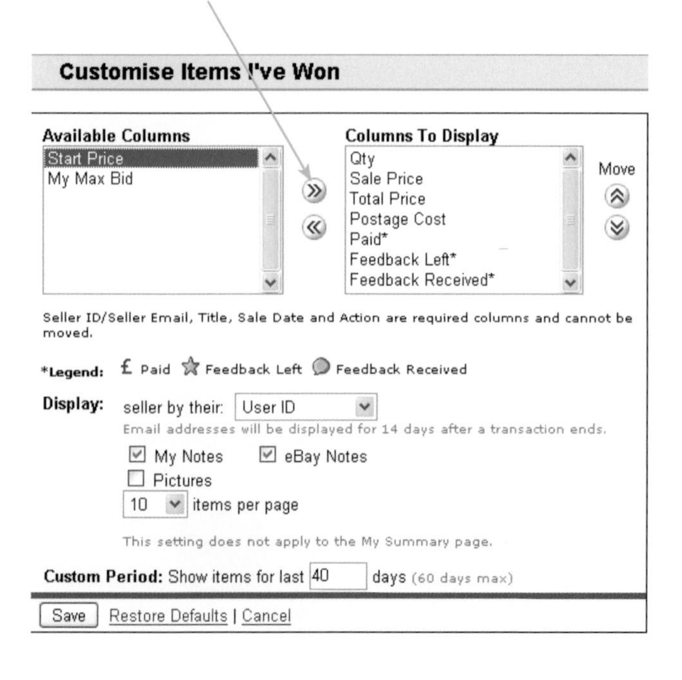

3. Click the buttons under Move to change the order in which the items appear

4. Click Save to save the customised settings

Save

Items I've Won (1 item)

Show: **All** | Awaiting Payment (0) | Awaiting Feedback (0)

	Seller ID	Qty	Sale Price	Total Price	Start Price
☐		1	£3.99	£6.44	£3.89

5. Return to the My eBay summary to view the customised items

Beware

If you include too many categories in the Items I've Won section it may start to look cluttered and difficult to follow. Decide which categories are the most important and stick to them.

My eBay favourites

As shown in Chapter One, you can create favourite searches for topics you want to check on regularly. These searches can be accessed from the My eBay page by clicking the All Favourites link:

All Favourites

- Searches
- Sellers
- Categories

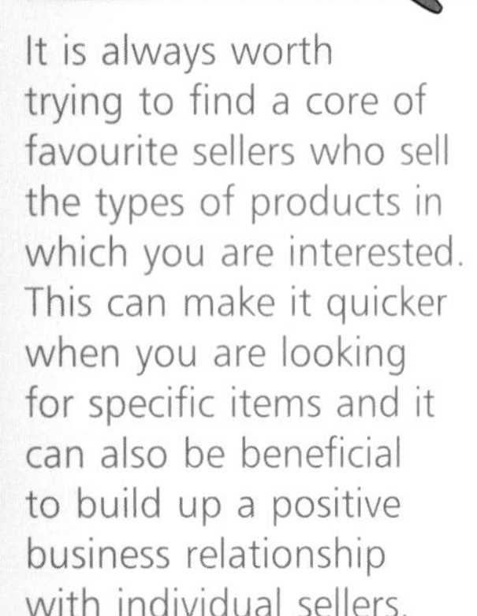

Hot tip

It is always worth trying to find a core of favourite sellers who sell the types of products in which you are interested. This can make it quicker when you are looking for specific items and it can also be beneficial to build up a positive business relationship with individual sellers.

In addition to favourite searches, it is also possible to create lists of favourite sellers and categories. If you find a seller you like, and you want to keep an eye on other items that they have coming up for sale, then you can add them as a favourite seller. For categories, you can select a specific category (a particular sub-category of the standard main categories) and view this over time as the items for sale within it change.

Adding favourite sellers

To create a list of favourite sellers:

1. Click on the Sellers link in the All Favourites box above

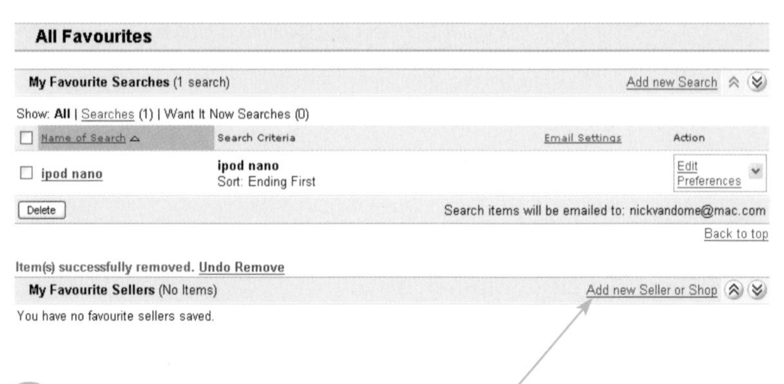

2. Click on the Add new Seller or Shop link in the All Favourites section of the My eBay summary

3. Enter the details of the seller or shop to be added to your list of favourite sellers

4. Click Continue

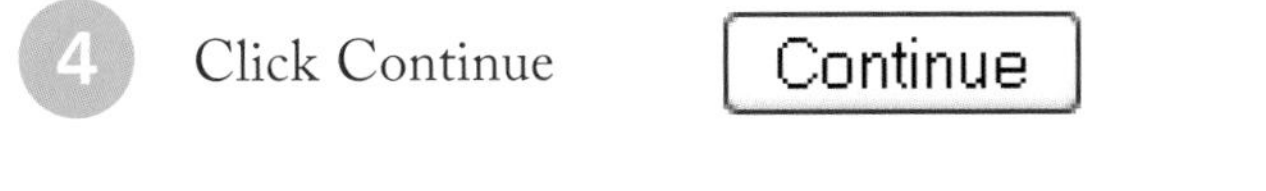

If you have selected a Power Seller as a Favourite Seller, make sure you check what they have for sale on a regular basis. Since Power Sellers usually rely on high volumes of sales, their inventory turns over at a rapid rate.

5. Click Add To Favourites

6. The seller is added to the All Favourites section within the My eBay summary. Click here to see the other items which they currently have for sale

My Favourite Sellers (1 seller)
Seller / Shop
Added
nickvan1 (18) me
View seller's other items
09-Feb-06

Adding favourite categories

To create a list of favourite categories:

1. Click on the Categories link in the All Favourites box in the My eBay summary

2. Click on the Add New Category link

...cont'd

3. Select a main category and drill down through the sub-categories until you reach the item you want

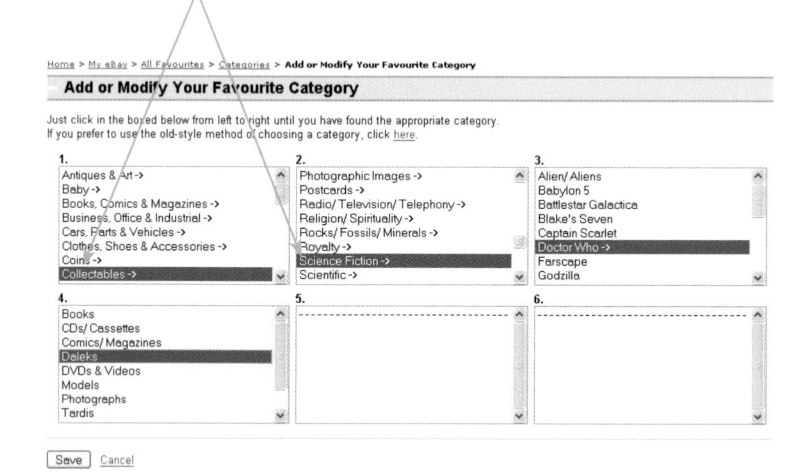

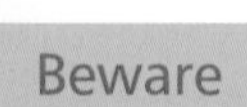
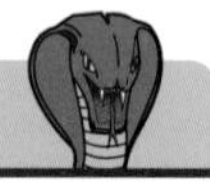

For every category that is selected, a category number should be shown in the box at the top of the window. If not, the category will not be included as a favourite.

4. Scroll to the bottom of the page and click Save to add the category as a favourite

5. The new favourite category can be viewed in the All Favourites section of the My eBay summary

6. Click here to view different options for the selected category, such as current auctions or upcoming ones

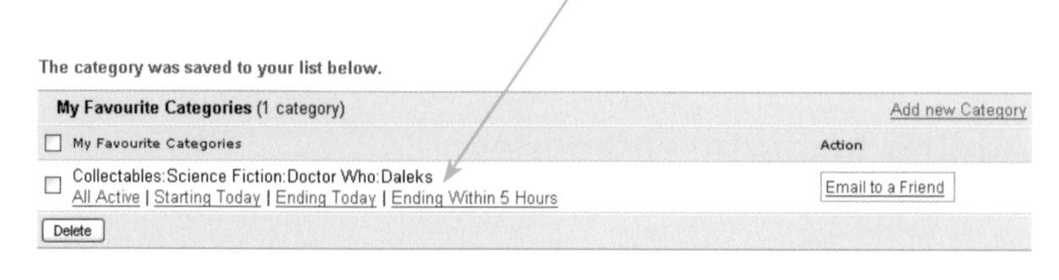

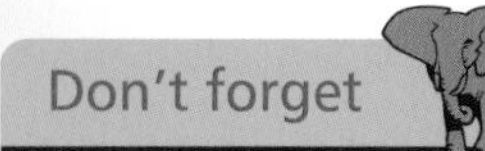

Up to four categories can be added to the Favourite Categories list.

Customising My eBay

Since the My eBay area is your own personal record of your activities on the site, it is only logical that you can customise it to appear the way you want. It is possible to add or remove sections of the My eBay page and also to change the order in which they appear. To customise a My eBay page:

1. Click here at the top of the summary page

Customise Summary

2. Select an item to add to the summary page

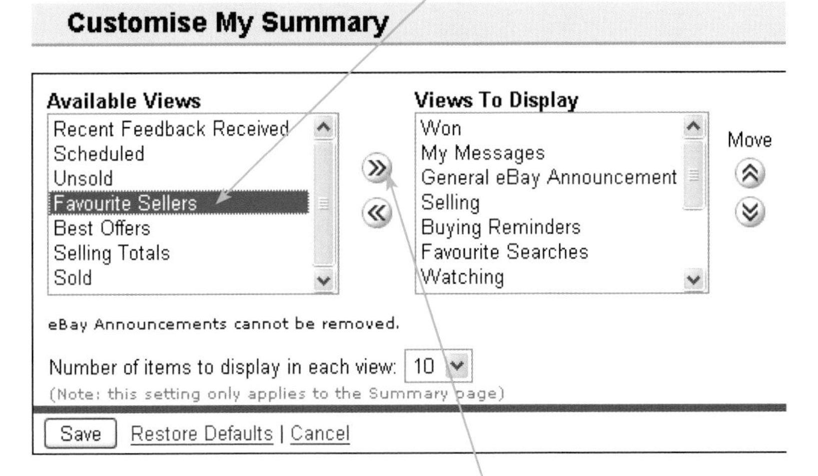

3. Click here to add the selected item

4. The selected item is added to the summary page

My Favourite Sellers (1 seller)

Seller / Shop

nickvan1 (18) me

View seller's other items

Hot tip

If you customise the My Summary page and do not like the end result, click on the Restore Defaults link in the Customise My Summary window to restore it to its original format.

...cont'd

Rearranging items

If you do not like the order in which items are presented on the My eBay summary page, for example if the one you access most frequently is at the bottom of the page and you have to scroll down to see it, then it is possible to rearrange the order in which they appear. To do this:

It can be beneficial to include elements with the fewest items at the top of the My Summary page. This will enable the items below to be visible without having to scroll down the page.

1. Once an item has been added to the list of items to be included on the summary page, select it in this panel

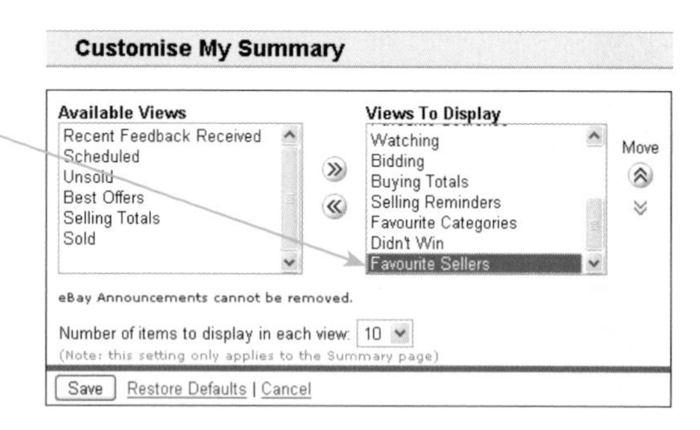

2. Click the buttons under Move to change the item's order. This will affect where it appears on the summary page

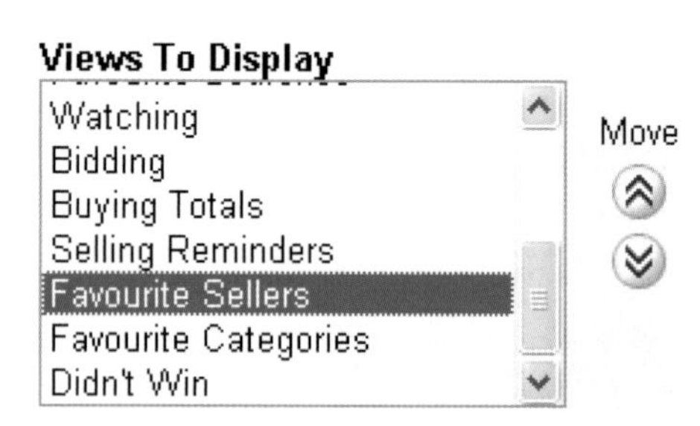

3. The item will appear at the selected point on the summary page

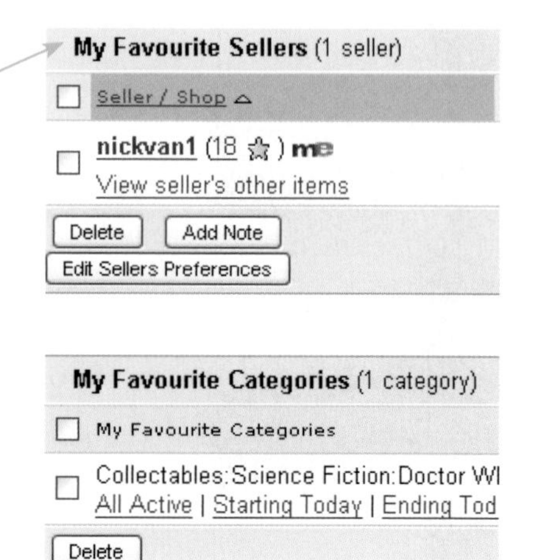

4 Beginning to buy

The thrill of the chase and the competitive element give auctions an added edge to merely buying items online. This chapter looks at how auctions work on eBay.

General buying guidelines

Once people have got to grips with the eBay site and the way it functions, the first thing they usually want to do is to start buying things. Even if you are looking at eBay as a source of income through selling, buying is definitely a good starting point. It is a good way of interacting with other eBayers and understanding more about the payment and feedback processes. However, before you start buying, there are some general guidelines that should be observed:

- Do some research about the item you want to buy; initially, this can be done on eBay. Resist the temptation to bid for the first similar item you come across when searching on eBay: for most searches you will find several items which could meet your needs. Initially, look at some of them to see the bid amounts they are attracting, what kind of condition they are in, the charge for the postage and packaging, how the seller would like to receive payment and where they are willing to send it. Compare this with other, similar items so that you can build up a good understanding of what is available. Once you have done this, add several similar items to your Items I'm Watching in My eBay and keep an eye on the auctions until they are finished. Then you will be able to see the final price that each item fetched and also the User ID of the person who has won the bid. This is particularly important if you think you will be bidding for similar items. If you keep a record of another eBayer's bidding activities it may be possible to second-guess how they are going to act if they are participating in the same auction as you at any stage.
- Do some research outside eBay. For a lot of items, such as cars and computers, there are trade magazines that offer prices for new and used items. Also, there are a variety of specialist magazines for collectable items that give detailed reports of how much certain items are worth, according to their condition and rarity. Before shopping on eBay it is worth looking at a guide related to the items in which you are interested.

A lot of traditional price guides are now being overtaken by eBay itself: the price paid on the site is usually a good reflection of how much an item is worth.

- There is little point in buying something on eBay if you can buy the same thing cheaper in one of your local shops. Before plunging into eBay, have a look around locally to see if similar items are available and, if so, the price at which they are sold.
- Establish whether you are happy with second-hand goods or if you want everything to be brand new. Some people love the thought of getting a bargain with a second-hand item, while others would never entertain the idea. This is a personal decision, and on eBay there are both new and used goods: check the description to see if an item is new or second-hand.
- Don't enter into an auction just for the sake of it. Sometimes when browsing through eBay you may look at something just out of interest or curiosity. Once the auction page is visible there can be a desire to place a bid just to try and win it. If possible, resist this and always ask yourself if you really want, or need, a particular item.
- Establish a maximum price you are prepared to pay for an item and stick to it. Once an auction starts it is easy to get carried away and let the competitive nature of the process develop into a 'win-at-all-costs' attitude. This is fine if you are determined to acquire an item but there is little point in winning if you pay too much as a result. For most items, similar auctions will be available soon.
- Avoid jumping in with a very high, early bid. Novice bidders frequently put in high bids as soon as they see an auction, in their desire to try and secure it. This not only puts them at a disadvantage (because other people can see that they are overly keen) but it is also unnecessary: if an auction has several days to go it is better to watch and see how the bidding is progressing before you enter the fray. If you put in a bid near to the closing time then there is less chance for other people to outbid you.

Don't forget

Contacting the seller is a good way to establish a rapport with other people in the eBay community and can help you feel more confident in your transactions.

Buy It Now option

Although eBay is known primarily as an auction site, there is also a facility for buying items immediately from a seller, without using the bidding process. This is know as Buy It Now (BIN). The Buy It Now option can be used in two ways: in conjunction with an auction, in which case the BIN option is removed once the first auction bid has been placed; or on its own, which results in a transaction that is the same as any non-auction online retailer.

Placing a starting bid on an auction that also has a Buy It Now price is known as BIN Stomping. This is done by buyers to remove the BIN price and hopefully acquire the item at a significantly lower price.

If it is used in conjunction with an auction, the BIN price is usually pitched at just over what the seller thinks they may get in the auction itself. For the buyer there could be several reasons to buy an item using the BIN option:

- They want the item as soon as possible and do not want to wait until the auction ends, which could be several days.
- They are happy with the price as it is less than they would have gone up to in the auction itself.
- They do not want the unpredictability of an auction.

If a Buy It Now sale does not also have an auction option, this is known as a Fixed Price format. This means that the item can only be bought at the stated price. To sell with this option you must have an overall feedback rating of 10 or higher, or a direct debit facility on file with eBay, or a PayPal account and an overall feedback rating of 5 or higher.

To buy an item using the Buy It Now option:

1. Locate an item that is on sale for a BIN price

DUNLOP - SPORT Squash Balls ..pack of (3).PSA

2. Buy It Now items are denoted by this icon

3. Click the Buy It Now button

Buy It Now price: £3.99

Buy It Now >

4. Check the purchase details and click Commit to Buy

Review and Commit to Buy

Hello nickvan1 (Not you?)

Item title:	DUNLOP - SPORT Squash Balls ..pack of (3).PSA
Buy It Now price:	**£3.99**
Postage and packing:	£2.45 -- Seller's Standard Rate. Additional services available.
Payment methods:	PayPal, Personal cheque, Postal Order or Banker's Draft.

By clicking on the button below, you commit to buy this item from the seller.

Commit to Buy

5. Click Pay Now to purchase the item

Buy It Now Confirmation

You bought this item

Pay Now > **or continue shopping with this seller**

Click **Pay Now** to confirm postage, get total price, and arrange payment

Hot tip

The Buy It Now option is useful for items such as presents, when you want something for a particular time and you do not want the uncertainty of an auction.

...cont'd

6. Select a payment option

Select a payment method (seller accepts the following)

(•) PayPal

You are eligible for up to £500 PayPal Buyer Protection coverage on this purchase.

( ) **Other accepted payment methods**
(Postal order/Banker's Draft; Personal cheque)

[Continue >]

Don't forget

The payment process for a Buy It Now item is the same as for an item won in an auction. For more details about paying for items, see Chapter Ten.

7. Click the Continue button to proceed to the payment section for the item

8. Alternatively, an email will be sent to you to alert you to the fact that you have purchased the item. Click the Pay Now button to pay for the item

Congratulations, the item is yours. Please pay now

Dear nickvan1,
Congratulations You committed to buy the following item:

DUNLOP - SPORT Squash Balls ..pack of (3).PSA

Sale price:	£3.99	
Quantity:	1	
Subtotal:	£3.99	
Postage	Seller's Standard Rate:	£2.45
	Royal Mail International Signed-for:	£5.88
Insurance:	(not offered)	

View item | Go to My eBay

Get Your Item

Click to confirm postage, get total price and arrange payment.

Multiple item auctions

Multiple item auctions, also known as Dutch auctions, are ones where a seller has several identical items for sale. Instead of selling them all in individual auctions they can choose to sell them as separate lots in a single auction. So if someone has 20 identical bicycles for sale they can list them in a single auction. The buyers can then put in a bid and also specify the quantity that they wish to buy. So someone could bid £50.00 and specify that they would like a quantity of three. If this is the final highest bid then they would pay £150.00 for the three bicycles (£50.00 each). In a multiple item auction, all of the highest winning bidders pay the same price. So in this example the next 17 highest bidders would also be able to purchase a bicycle for £50.00.

Multiple item auctions work in a similar way to single item auctions, with the exception that the current high bid is determined by the price offered multiplied by the quantity ordered. So someone bidding £10.00 and asking for a quantity of three would have a higher bid than someone bidding £25.00 and asking for a quantity of one.

Although multiple item auctions are a perfectly legitimate part of eBay, they do not always find great support from sellers. This is because they have to sell all of the items at the same price. A more common occurrence is for sellers to list multiple items with just a Buy It Now price. Buyers can then buy one or more of the items at a set price.

1. Multiple item auctions at a set Buy It Now price are now a regular feature on eBay

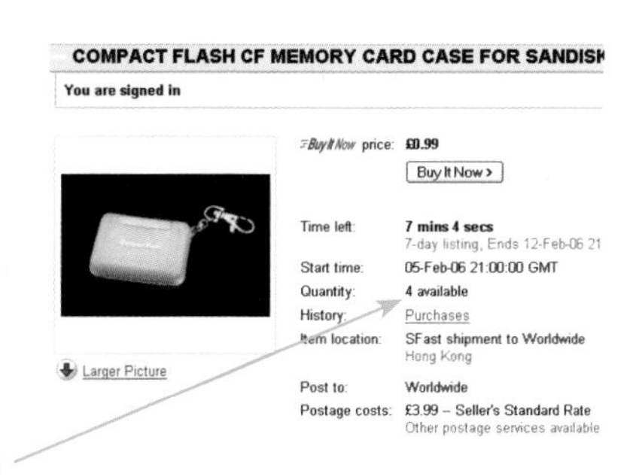

2. The number of items available is listed here and buyers can purchase one or more

How an auction works

An auction on eBay is similar in some ways to one held in an auction house. However, since eBay auctions operate in an online environment, there are two important differences.

The first difference with an eBay auction is that users can place bids that are higher than the current price of an item (typically the top price that they are prepared to pay) and eBay will automatically bid up to this maximum price. For example, if the current high bid is £1 and someone then bids a maximum of £10, the current high bid moves up to £1.20. If no-one else bids on the item then the bidder will get it for this price, rather than the £10 maximum. However, if someone else bids, eBay will keep raising the current high bid up to the maximum bid by the first bidder. So if a second bidder comes in with a maximum bid of £3, eBay will raise the current high bid to £3.20 for the first bidder. To become the high bidder, the second bidder will have to put in a bid of over £10, i.e. the first bidder's maximum price. This form of bidding is known as proxy bidding, which means that eBay will bid automatically on your behalf up to the maximum you have stated when you first bid. When a proxy bid is placed it goes up by a specific incremental level (see following page for details).

You never see the maximum amount that someone has bid, only the current high bid. In some cases this may be the same, but not necessarily.

The second difference with an eBay auction is that it is for a set duration. Currently, the maximum duration for an auction is 10 days and the minimum is 1 day. The duration of an auction is set by the seller at the time of listing the item. Some points to consider when looking at the duration of an auction are:

- If the duration is too short then people may miss the auction
- If there are several days to go for an auction there is no need to place a bid immediately. Wait and see how the auction progresses and then bid accordingly.
- If you are unavailable when the auction finishes you can place a maximum bid and let eBay place proxy bids.

Hot tip 

When looking at the duration of an auction for an item in which you are interested, try to make sure you will be around at the end of the auction. This will enable you to make a last minute bid, if required.

Bidding increments

The amount by which a high bid increases once someone bids above it is set by eBay. The current incremental bid levels are:

- If the current high bid is between £0.01 and £1.00 the increment level is £0.05 (so that, if the current high bid is £1 and someone bids a maximum of £5, the current high bid would go up to £1.05)
- If the current high bid is between £1.01 and £5.00 the incremental level is £0.20
- If the current high bid is between £5.01 and £15.00 the incremental level is £0.50
- If the current high bid is between £15.01 and £60.00 the incremental level is £1.00
- If the current high bid is between £60.01 and £150.00 the incremental level is £2.00
- If the current high bid is between £150.01 and £300.00 the incremental level is £5.00
- If the current high bid is between £300.01 and £600.00 the incremental level is £10.00
- If the current high bid is between £600.01 and £1,500.00 the incremental level is £20.00
- If the current high bid is between £1,500.01 and £3,000.00 the incremental level is £50.00
- If the current high bid is over £3,000.01 the incremental level is £100.00

Since the incremental levels are dependant on the current high bid, these can change as an auction progresses e.g. if the high bid is £10 the incremental level will be more than if it was £2.

Prices given in this book were correct at time of going to press.

Don't forget 

Bidding increments determine the minimum amount someone has to bid over the current high bid to become the new high bidder. If you try to put in a bid that is under the next incremental level, eBay will not allow you to do so.

Example auction

When you start bidding in auctions the proxy system and incremental levels can be a bit confusing. Watch a few auctions first and start by bidding on auctions for low-priced items. The following example shows how an auction operates using the proxy system and bidding incremental levels:

- A seller lists an item for sale at £1. This is known as the starting price and is the minimum price someone has to bid to become the high bidder and have the current high bid.
- Bidder A makes a bid of £1, known as the starting bid. The current high bid is therefore £1.
- Bidder B wants to make a bid and sees that they have to bid a minimum of £1.05 (the next incremental level up for an item at this price). Bidder B does this, and becomes the high bidder with a current high bid of £1.05.
- Bidder C wants to up the bidding. They see the next bid is £1.25 or more (since the item has gone into the next incremental level of £0.20). However, they want to make a more significant bid and so bid a maximum of £16.00. They become the current high bidder and the current high bid is £1.25: the next incremental level above the previous high bid.
- Bidder B sees that they have been outbid and that the next bid has to be £1.45 or higher. They place a bid of £1.50 but receive a message to say that they have been outbid. This is because Bidder C's maximum has not been reached, so eBay places a proxy bid on their behalf. This raises the current high bid to £1.70: Bidder B's bid of £1.50 plus the next incremental level of £0.20. Bidder C is still the current high bidder.
- Bidder B still wants the item and sees that the current high bid has to be £1.90 or higher (the current high bid plus the next increment level of £0.20). They decide to place a bid of £6.00, but again receive a message to

Don't forget

Some sellers set a reserve price for an item. This is a minimum amount which they are prepared to accept for an item. If the reserve is not met then they do not have to sell. There are advantages and disadvantages in having a reserve price and this is looked at in more detail in Chapter Seven.

Don't forget

When bidding for an item, you cannot see someone else's maximum bid, only the next incremental level at the current price.

say that they have been outbid. Once more, eBay has placed a proxy bid for Bidder B, since Bidder C's bid has not yet matched Bidder B's maximum bid. So the current high bid becomes Bidder C's bid plus the next incremental level (which has now risen to £0.50 as the current high bid is over £5.01). So the new current high bid is £6.00 plus £0.50, making £6.50.

- At this point Bidder B decides to drop out of the auction. Although they can see the current high bid, they do not know Bidder C's maximum bid (as this is not displayed). They therefore decide that they are not prepared to pay more.
- Bidder A now re-enters the auction. They now realise that Bidder C is an experienced operator and has probably put in a fairly high maximum bid. Bidder A therefore puts in a bid of £19.00. This surpasses Bidder C's maximum bid of £16.00 and so Bidder A becomes the current high bidder. The current high bids becomes £17.00, which is Bidder C's maximum plus the increment at this price point of £1.00. (Bidder A could have become the current high bidder with a bid of £16.01 but they had no way of knowing Bidder C's maximum bid.)
- Bidder C now has a difficult decision to make. They know that Bidder A is a serious bidder and they have to decide how badly they really want the item and how much they are prepared to spend. The current amount they would have to bid is £18.00: the current high bid plus the incremental amount of £1.00. Bidder C places a bid of £21.00, making the current high bid £20.00: Bidder A's maximum bid of £19.00 plus the £1.00 increment.
- Bidder A realises that they have serious competition. Because of this, they wait until the last few minutes of the auction and put in a final bid of £25.00. They win the auction and the final price is £22.00: Bidder C's final maximum bid of £21.00 plus the £1.00 increment.

Beware

Although it can be frustrating to be outbid in an auction, don't get into a bidding war just to beat someone else. You may end up winning the auction, but at an inflated price.

Hot tip

It is perfectly acceptable to watch an auction for several days and then come in with a bid at the last moment. Bidding strategies like this are looked at in more detail in Chapter Five.

Bidding history

Hot tip

Looking at the bidding history can be a good way of getting to know the habits of other people in the auction. For instance, someone might bid at the same time each day; someone else might make a number of consecutive bids in order to become the current high bidder. This can be useful information when it comes to bidding tactics, particularly towards the end of an auction.

Don't forget

Whenever you put in a bid it will be registered in the bidding history, even if you do not become the high bidder. This is because you can automatically be outbid if someone else has put in a higher maximum. So, in effect, your bid would be the high bidder for a split second until the proxy bid is placed by eBay. Therefore it is possible to have several bids appear consecutively in the bidding history, without becoming the high bidder with any of them.

There are certain details that can be ascertained from watching an auction. One of these is the bidding history for a particular item. This can be useful as it shows who has been bidding and how many bids they have had to make to become, or try to become, the highest bidder. To view the bidding history:

1. Open the auction page and click here

Time left:	**1 hour 51 mins** 5-day listing, Ends 12-Feb-06
Start time:	07-Feb-06 22:47:37 GMT
History:	38 bids (£0.99 starting bid)
High bidder:	

2. The current highest bidder is displayed at the top of the list of bidders. The date displayed is when the current high bid was placed. The high bid is not necessarily the maximum bid by the high bidder, just the maximum required to beat any other current bids

User ID:	Bid Amount	Date of bid
	£234.00	12-Feb-06 14:56:15 GMT
(94 ★)	£229.00	12-Feb-06 02:50:11 GMT
(7)	£225.00	12-Feb-06 14:55:52 GMT
(7)	£215.00	12-Feb-06 14:55:26 GMT
(7)	£205.00	12-Feb-06 14:55:02 GMT
(0)	£195.00	12-Feb-06 02:07:52 GMT
(94 ★)	£172.00	11-Feb-06 04:30:01 GMT
(94 ★)	£170.00	11-Feb-06 04:29:28 GMT
(0)	£160.50	12-Feb-06 02:07:35 GMT
(94 ★)	£150.00	11-Feb-06 04:29:18 GMT
(2)	£140.00	10-Feb-06 21:18:17 GMT

3. All bids that did not beat the high bidder are shown below it. Look at the times to see when these bids were made. In some cases, several bids can be made one after another but they still do not beat the highest bid, so the bidder tries again

5 Bidding in an auction

The majority of sales on eBay are done through auctions. This chapter looks at the bidding process and shows how to develop a strategy to give yourself as good a chance as possible of winning items.

Bidding for an item

Once you have done all of your research, and understand fully the way in which the bidding works, it is time to start participating in an auction. This is the essence of eBay and it can be an exciting and stimulating experience. But be careful not to get too carried away. To start bidding for an item:

Hot tip

Start by bidding for some lower-priced items, to give yourself a chance to get to know how the system works.

1. Find the item in which you are interested and open the auction page. Check the details and, if you are happy with them, click the Place Bid button to start bidding for the item

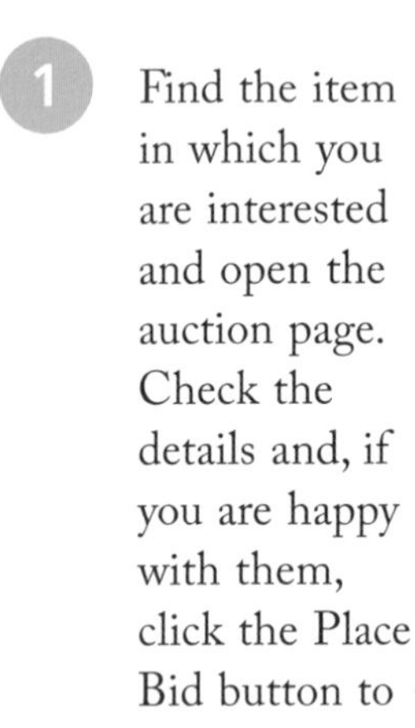

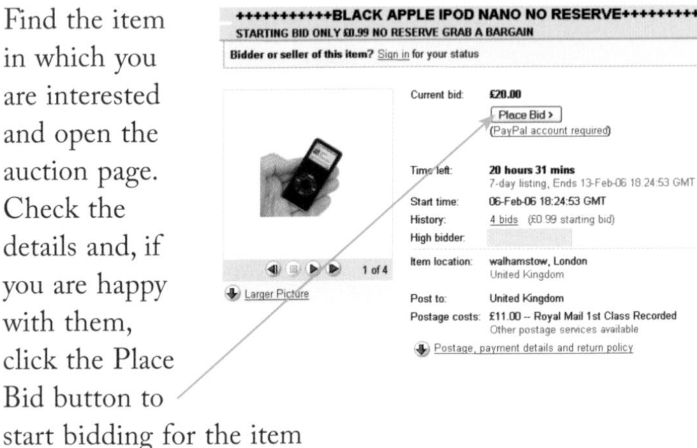

2. The amount of the current high bid is shown at the top of the page. To become the high bidder you have to beat this price by at least the next incremental step in the auction

Don't forget

Matching the next incremental level shown on the Place a Bid page does not guarantee you will become the current high bidder. This depends on the maximum amount bid by other users.

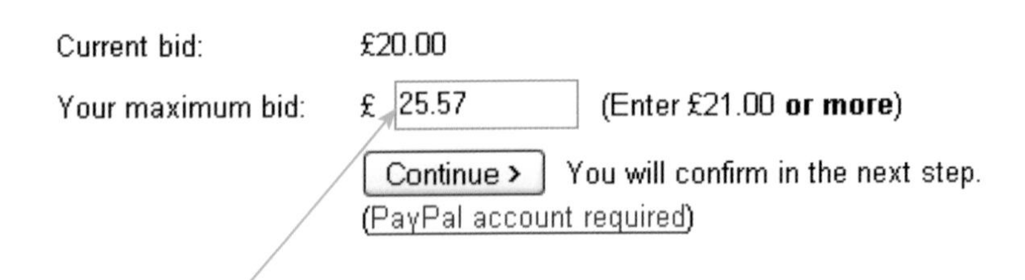

3. Enter the maximum amount that you are prepared to bid at this point. The minimum amount required is shown to the right of the bid amount box. Click the Continue button to move to the next step of the bidding process

4. Check the details of your bid carefully

Review and Confirm Bid

Hello nickvan1 (Not you?)

Item title:	+++++++++++BLACK APPLE IPOD NANO NO RESERVE++++++++++++
Your maximum bid:	**£25.57**
Postage and packing:	£11.00 -- Royal Mail 1st Class Recorded. Additional services available.
Payment methods:	PayPal

By clicking on the button below, you commit to buy this item from the seller if you're the winning bidder.

Confirm Bid

5. Click the Confirm Bid button to enter your bid

Confirm Bid

6. If your bid is above the next incremental level of the current high bid then you will become the current high bidder. If this is the case, a Bid Confirmation box appears on the auction page

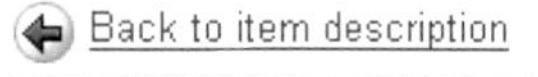

Bid Confirmation

You are signed in

Important: You are one bid away from being outbid. If another user places a bid, you will not win. To increase your chances of winning, enter your highest maximum bid.

Title: +++++++++++BLACK APPLE IPOD NANO NO RESERVE++++++++++++

Time left:	20 hours 28 mins
History:	5 bids
Current bid:	£25.57
Your maximum bid:	**£25.57**

Enter Higher Maximum Bid — Increase your chances of winning

or Check your status — See if you're still the high bidder

Beware

Make sure you review your bid carefully and that you have typed the correct amount. Mistakes can be rectified later but it is better to get it right first time.

Don't forget

Once a bid has been made and confirmed, there is a binding legal contract between the buyer and the seller.

...cont'd

7. If your bid is not higher than the current high bidder's maximum, the following page will be displayed. Click the Bid Again button if you want to submit another bid

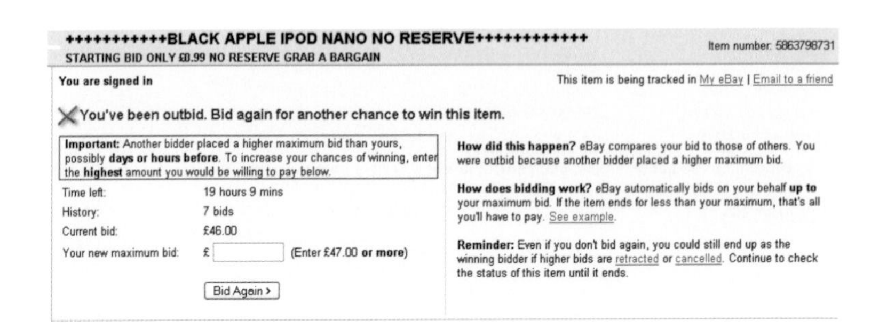

8. Once you have become the high bidder, the current high bid is displayed (this is not necessarily your maximum bid)

Current bid: **£25.57**

Place Bid >

(PayPal account required)

Time left: **20 hours 22 mins**
7-day listing, Ends 13-Feb-06 18:24:53 GMT

Start time: 06-Feb-06 18:24:53 GMT

History: 5 bids (£0.99 starting bid)

High bidder: nickvan1 (18 ☆) me

9. Your User ID is displayed here as the high bidder

Beware

If you see that you have been outbid by a proxy bid (a bid made automatically by eBay because someone else has put in a higher maximum bid), review your own bidding strategy, particularly in light of the maximum you are willing to pay for the item. Once the price reaches your own limit, stop bidding. You can always check the final price at the end of the auction by visiting the auction page.

Viewing high bidder details

Once you have become the high bidder for an item, you can check in a number of ways that you haven't been outbid:

1. Click on the Bid History link on the auction home page. Your name should be displayed at the top of the list

User ID:	Bid Amount	Date of bid
nickvan1 (18 ☆) me	£25.57	12-Feb-06 21:56:39 GMT
	£25.00	12-Feb-06 17:03:37 GMT
	£19.00	12-Feb-06 17:04:59 GMT

2. You will be sent an email telling you about your high bidder status and giving details of the item

Bid confirmed. You are the high bidder!

Dear nickvan1,
You are currently the high bidder for the following eBay item

++++++++++BLACK APPLE IPOD NANO NO RESERVE++++++++++++
Your current bid: £25.57
Your maximum bid: £25.57
View item | Go to My eBay

3. The item will be added to your My eBay page, under the Items I'm Bidding On section

Hot tip

Even when you are the current high bidder, it is a good idea to check the bid history to see what bidding activity has gone on and who is involved in the auction.

The price for items that you are winning are shown in green on the My eBay page. Items for which you have currently been outbid show the price in red.

Being outbid

Even if you are the current highest bidder for an item, you can still be outbid later by another user. This just means that someone else will have put in a bid which is higher than the maximum figure you entered.

If this happens, you are notified about it in two ways:

1. On the auction page, a notice appears to alert you to the fact that you have been outbid. Click Bid Again to place another bid

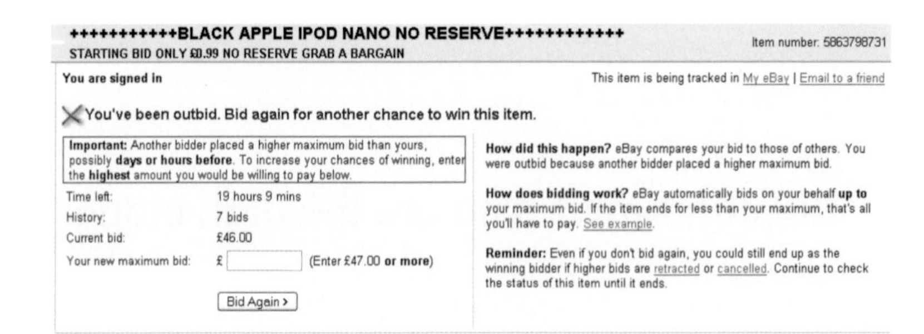

2. An email is sent to you detailing the fact that you have been outbid. Click Bid Again to place another bid

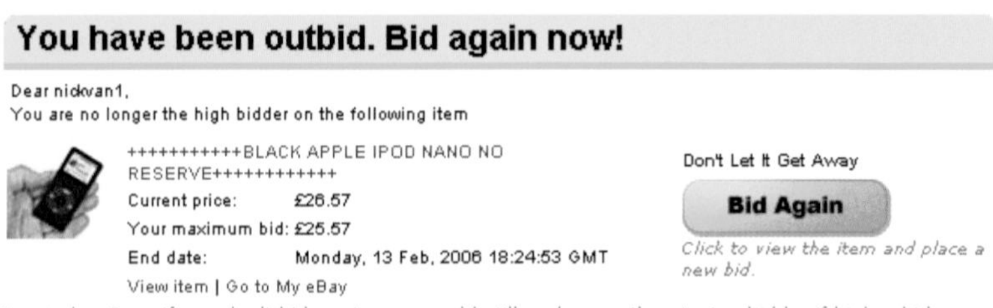

eBay always makes it as easy as possible for you to place another bid once you have been outbid.

Asking questions

During an auction there may be questions that you would like to ask the seller. These could involve the condition of the item, the postage details, or a specific question about something in the description of the item. If there is anything that you want to ask it is essential that you do so: after the auction has ended it will be too late. When you receive an answer to your question, if there is anything that you are still unsure about then ask another question. There is no limit to the number of questions you can ask a seller, but apply some common sense when doing so and do not bombard them with unnecessary queries. To ask a seller a question:

Ask questions before you place a bid rather than afterwards, when it may be too late.

1. Access the auction page and click the 'Ask seller a question' link in the Seller Information box

Seller information

Feedback Score: 4
Positive Feedback: 100%
Member since 03-Feb-06 in United Kingdom
Registered as a private seller
Read feedback comments
Add to Favourite Sellers
Ask seller a question
View seller's other items

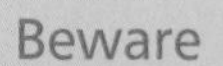

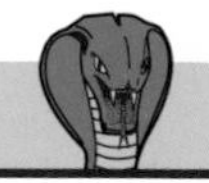

If you do not get an answer to a question then you should not proceed with bidding for the item. This could be an indication that there is a problem with this particular seller.

2. Enter your question here and click the Send button

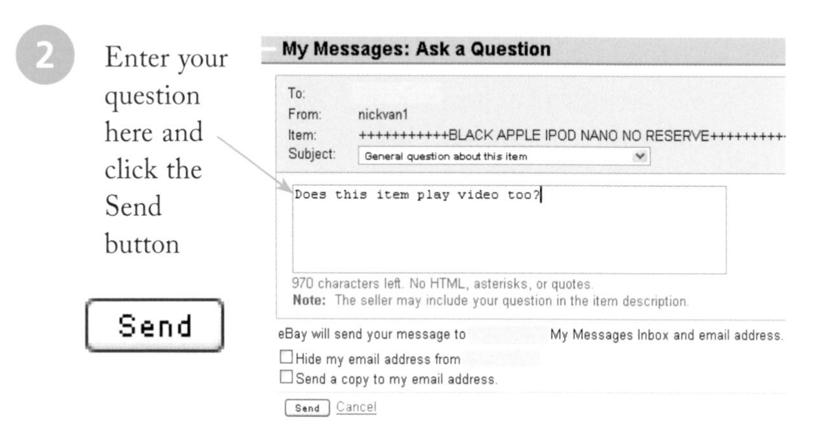

Bidding to win

No one likes losing an auction, particularly if it is for an item they really want. With experience you will get to develop your own bidding strategies, but some simple tactics can give you a significant advantage over other bidders:

- Don't jump in too early with a high bid. When you first view an auction, look to see when it finishes. If it is several days until the end (and you know you will be around to watch the auction) there is no need to place a bid. Instead, just watch the auction to see what activity is taking place. See who else is bidding and by how much the current high bid is increasing.
- Test the water. Once you have watched an auction for a few days place an initial bid to see what response it gets. Then watch to see if any higher bids come in. If they do, then you know you have competition. At this point it is best to watch the auction again for a while: there is no point in getting into a bidding war at an early stage, as this could enable someone else to come in with a higher bid once you have reached your maximum bid. Ideally, you want to be able to come in with your highest bid as close to the end of the auction as possible.
- Enter uneven numerical values. One mistake that a lot of novice eBay bidders make is to bid in round figures – £5.00, £10.00, £200.00, etc. While this may seem logical, it makes it easier for more experienced eBayers to win the auction by only a few pence. For instance, if someone bids £5.03 rather than £5.00 then they could become the current high bidder because of the extra three pence. Always add odd figures like this to the end of bids, as it increases your chances of outbidding people who use round figures.
- Check out the opposition. Watch the bid history of an item to see if other bidders bid at a particular time. For instance, someone may bid in the evening, whereas someone else may bid during the afternoon. If people bid near the end of an auction then you know that they will probably be watching it right up to the closing time.

If another user bids back immediately after you have placed a bid then it is a sign that they are keen on the item. If there are a few days left for the auction, you do not need to place another bid immediately.

- Make sure you are in a position to bid towards the end of the auction. If you know you are not going to be near your own computer when the auction is due to finish, try to make alternative arrangements: if there is a friend or relation who could place a bid on your behalf then give them details of what to do and how much to bid. Otherwise try and access eBay from another computer.
- Refresh your browser regularly when watching an auction. The time left is only updated if you refresh your browser window; it does not happen automatically.
- Enter your maximum bid towards the end of the auction. If you have been following an auction and it has still not reached the maximum which you are prepared to pay then you can enter this amount with hours, or preferably minutes, left on the auction. If you become the high bidder it does not mean that you will necessarily have to pay your maximum amount, as the current high bid will go up to the next increment above the previous high bid. There is no point in putting your maximum bid in at the beginning of an auction, as this just gives other bidders a target to aim for and several days to do so. Bidding at the last moment in an auction is known as 'sniping' and this is looked at on page 82.
- Don't panic if you get into a bidding war with someone else. This can be one of the most exciting aspects of eBay but it is important not to get carried away. Don't keep bidding just for the sake of it: if the price goes above the maximum you have decided on, then stop bidding (although it can be worth going over it by a small amount, just in case). If you want to keep bidding, don't enter a bid immediately after someone else has outbid you. Evaluate the bid that has been made and check the bid history to see how often the current high bidder has bid. Also, check the time left for the auction: if it is still a few days then you can happily stop bidding at this point and wait until the auction is nearer the end. A cool head is often successful when bidding on eBay.

Beware 

Don't be tempted to increase the maximum you are prepared to pay for an item just because you have been outbid.

...cont'd

View the bidding history of other users

If you find you are frequently bidding against the same people (which can happen if you are buying the same types of items) then it is a good idea to try and find out as much as possible about their bidding habits. To do this:

1. Access the Advanced Search page and click here to search for details about a specific eBay member. You will have to know their User ID in order to continue with the search

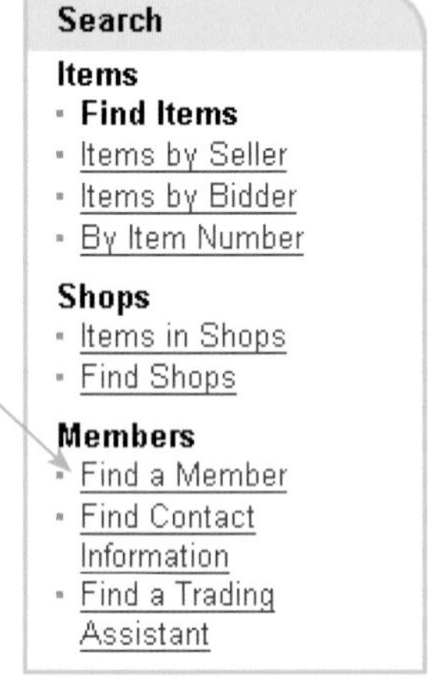

2. Enter the User ID and select the Feedback Profile option. Click Search

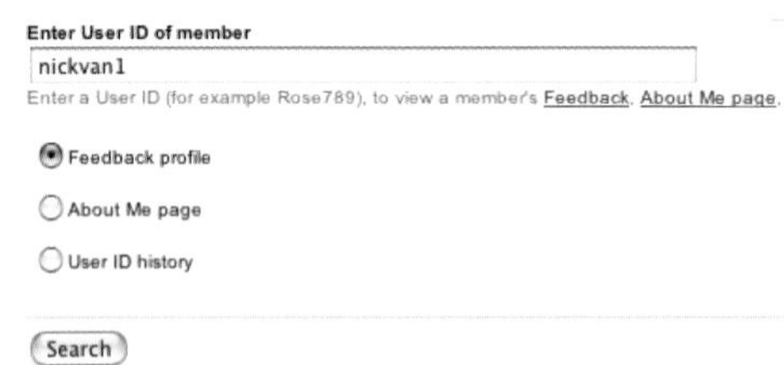

3. From the user's feedback, click on a link to view items that they have won

Hot tip

Other eBayers will probably be checking your bidding history too. Try to keep them on their toes by bidding at irregular times and try to avoid creating consistent bidding patterns.

4. Click here to view the bid history of the item

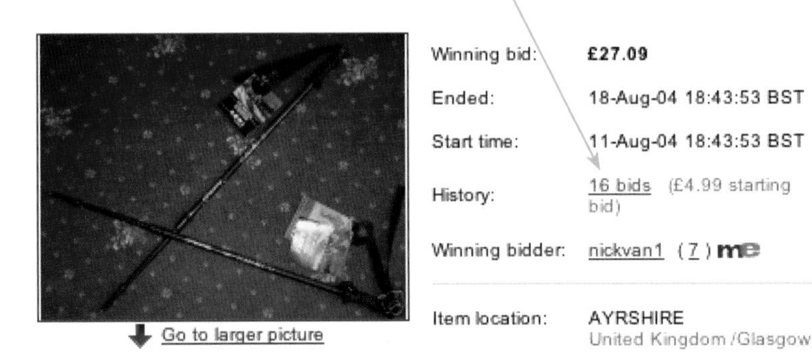

5. Study the bid history for the auction that has ended. The final bid amount does not necessarily relate to the maximum bid that was entered. The important item is the Date of Bid column. Check to see the time of the final bid and compare this with the closing time of the auction (see previous image). This will give you an idea of the user's bidding habits and how closely they watch auctions

Time left: Auction has ended.

Only actual bids (not automatic bids generated up to a bidder's maximum) are shown. Automatic bids may before a listing ends. Learn more about bidding.

User ID	Bid Amount	Date of bid
nickvan1 (7) me	£27.09	18-Aug-04 18:41:25 BST
	£26.09	18-Aug-04 16:34:02 BST
nickvan1 (7) me	£25.09	18-Aug-04 15:50:26 BST
	£25.00	15-Aug-04 22:40:34 BST
nickvan1 (7) me	£23.09	18-Aug-04 15:50:09 BST
nickvan1 (7) me	£21.09	18-Aug-04 15:49:43 BST
	£19.00	17-Aug-04 18:52:46 BST
	£14.50	16-Aug-04 16:35:35 BST
	£13.00	16-Aug-04 16:35:02 BST
	£11.00	15-Aug-04 17:08:31 BST
	£10.00	12-Aug-04 20:11:08 BST
	£10.00	15-Aug-04 17:08:22 BST
	£9.00	15-Aug-04 17:08:07 BST
	£8.00	15-Aug-04 17:07:52 BST
	£6.20	14-Aug-04 19:30:18 BST
	£5.19	14-Aug-04 19:29:55 BST

Beware

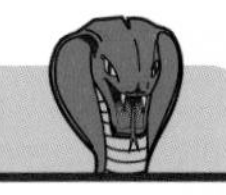

Some buyers always put in their first bid when there are only a few seconds of an auction left. If you find you are up against someone like this you will have to try to beat them at their own game by timing your bid even more close to the end of the auction.

Sniping

The best way to try to win an auction is to put in a bid at the last moment and hope that it is high enough to become the current high bid. If this is the case, it will not give other bidders time to respond. In eBay terminology this is known as 'sniping'. Although some sellers do not like this practice, as they feel it means buyers do not make as many bids against each other as they could, it is perfectly legitimate and allowable within the eBay rules and regulations. It is also great fun.

When participating in sniping, remember:

- A fast Internet connection will enable you to leave your bid closer to the end of an auction.
- If you are sniping on an auction then other people may be doing the same. Always be prepared to be out-sniped.

If you want to make a bid at the very last moment, keep one browser window open with the current high bid and time left displayed. Then open another window and enter a final bid. Keep this ready at the submit stage until the last moment. You can still view the current bidding activity and time left, and then send your final bid just before the auction closes:

Don't forget

A broadband Internet connection should enable you to put in a bid in the last few seconds of an auction. With a dial-up connection you may want to give it a couple of minutes. However, with both methods there is a chance of missing the end of the auction if there is a lot of Internet traffic at that time.

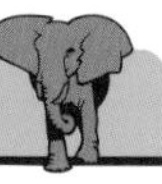

Don't forget

If you are going to be sniping manually, it is important that you are at your computer, or can access eBay, at the time when an auction in which you are interested is due to finish.

1. By keeping two browser windows open for the same auction you can have a bid ready and also view the latest high bid

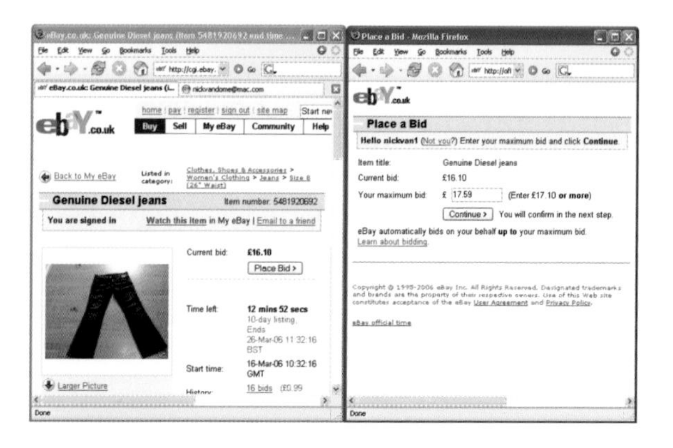

Automated sniping

Rather than having to sit at your computer and wait for the end of every auction in which you are interested, just so you can put in a snipe bid, it is also possible to use an automated sniping tool to do the work for you. These are online services that will place a bid for you on eBay at a time predetermined by you. After that you can turn off your computer, sit back, put your feet up and wait and see if the sniping software has won the item for you. There are a few points to bear in mind when using sniping software:

- As with most things connected with computers, sniping software does not always work perfectly. Bids can be missed, or made at the wrong time, although this is a very rare event.
- There is a fee for most sniping software services, usually based on the amount of use or at a flat fee per month.

Some online sniping service include:

- BidRobot at **www.bidrobot.com**
- Cricketsniper at **www.cricketsniper.com**
- AuctionSniper at **www.auctionsniper.com**
- AuctionBytes at **www.auctionbytes.com** (this contains an overview of most sniping sites and links to them)

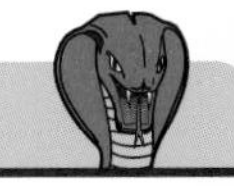
Don't forget

Although automated sniping software takes some of the fun out of bidding, it can be a very effective way of winning auctions.

Beware

Even automated sniping tools are subject to delays due to Internet traffic, which could have an impact on a bid that is timed to be placed a few seconds before the end of an auction.

Cancelling a bid

Except for a handful of circumstances, a bid on eBay is a legal contract. This means that the buyer is obligated to pay the seller the price of the winning bid and the seller has to provide the buyer with the item in the auction. In general, eBay frowns upon people cancelling their bids and will only allow it in certain circumstances. Some of the reasons for which eBay will not allow a bid to be cancelled is if a buyer decides that they do not really want an item once they have bid for it, or if they find the same item for cheaper elsewhere. The circumstances under which eBay will allow a bid to be cancelled are:

- If you mistakenly enter the wrong amount for a bid; for example if you enter £100.00 instead of £10.00.
- The description of an item does not match the product.
- You cannot reach the seller once the auction has finished.

To cancel a bid:

Beware

Before you cancel a bid due to lack of contact with the seller, you have to make a reasonable effort contacting them – something more than one unanswered telephone call. If in doubt, contact eBay and ask their advice.

1. Open the auction page and click here to view the bidding history

Current bid:	**£30.09**
	Place Bid >
Time left:	**5 hours 36 mins** 1-day listing, Ends 27-Mar-06 BST
Start time:	26-Mar-06 21:24:01 BST
History:	8 bids (£5.00 starting bid)
High bidder:	

2. Click here at the bottom of the Bid History page

User ID:	Bid Amount
	£30.09
	£30.00
	£27.00
	£22.00
	£20.00
	£18.00
	£16.00
	£6.00

See how to cancel bids if you need to.

3. Read the information about bid cancellation carefully. This should be seen as a last resort and, even though it is relatively easy to do in the right circumstances, it is not something that you should abuse or overuse. Even if you manage to cancel your bid there is always a chance that someone might give you negative feedback as a result. Click here to access the Bid Cancellation form

Cancel Bids

In general, sellers should not cancel bids on their auctions.

Legitimate reasons to cancel a bid include:

- A bidder contacts you to back out of the bid.
- You cannot verify the identity of the bidder after trying all reasonable means of contact.

However, there are tools in place to let you cancel bids.

Please note that bids **can't be reinstated** once they've been canceled.

Note: You will also be given a choice to cancel bids when you end your listing early.

Don't forget

Always double-check the amount that you enter when you place a bid. This will reduce the possibility of having to cancel the bid as the result of a mistake or mistyping.

...cont'd

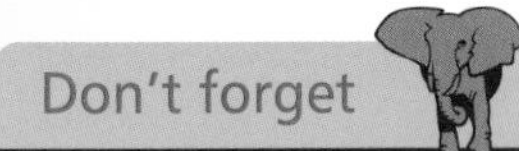

The Item Number is displayed on the main auction page for each item listed.

4. Enter the number of the item for which you placed a bid and select a reason for the cancellation

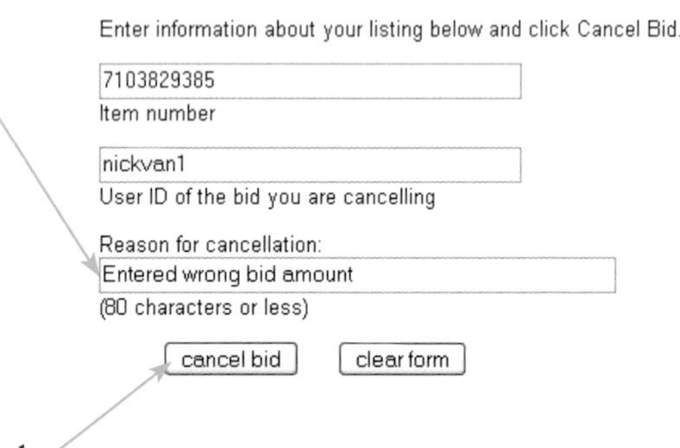

Enter information about your listing below and click Cancel Bid.
7103829385
Item number
nickvan1
User ID of the bid you are cancelling
Reason for cancellation:
Entered wrong bid amount
(80 characters or less)
cancel bid
clear form

5. Click the Cancel Bid button

Bid cancellation is annoying not only for the seller but also for other people who may have bid in the same auction. This is because your retraction could alter the dynamics of the whole auction.

6. All of the bids that you have placed for the item are cancelled, not just the one that was made in error. Confirmation of the bid cancellation appears after the Cancel Bid button is clicked

7. Once a bid has been cancelled, these details are displayed at the bottom of the Bid History page, for everyone to see

Bid retraction and cancellation history

User ID	Action / Explanation
nickvan1 (6)	**Retracted:** £33.90 **Explanation:**Entered wrong bid amount

Bid cancellation details also appear on your own User ID profile page.

8. If a bid is cancelled because of entering the wrong amount you have to go back to the auction and enter the bid that you intended to place initially. If you don't you will incur the displeasure of eBay and perhaps end up with negative feedback from the seller

Winning an auction

A lot of experienced eBayers leave their bidding until the final hours or minutes of an auction. This gives them a chance to see the pattern of the bidding and who is involved in the auction. For this reason it is useful to be able to view an auction in which you are participating up until it closes. If you are determined to win an auction then some proactive monitoring is required:

1. Open the auction page and check how long there is left for the auction. Also check that you are still the high bidder

2. Update the auction page at regular intervals by refreshing your browser. Ideally, this should be done until the final seconds

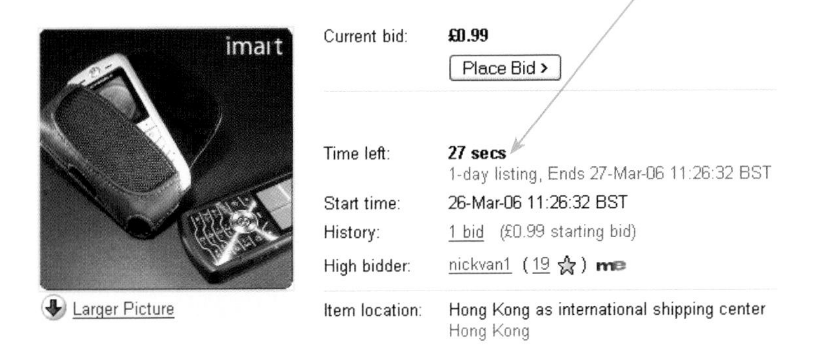

Hot tip

If you know that you will not have access to your own computer at the time an auction ends, try to make sure you can get access to the Internet through another computer. This could be done from work, if you are permitted to do this, or from an Internet café.

Don't forget

If you are participating in multiple auctions at the same time, open different windows in your browser so that you can view the auction home pages for each item simultaneously.

...cont'd

3. If you are still the high bidder when the auction closes the following message will appear on the auction page

Motorola SLVR L7 L6 V8 Leather Case

You won the item!

Pay Now > **or continue shopping with this seller**

Payment methods and related issues are looked at in more detail in Chapter Ten.

4. Click the Pay Now button to progress to the payment section

5. You will also be sent an email telling you that you have won the item. Click the Pay Now button to progress to the payment section

Congratulations, the item is yours. Please pay now

Dear nickvan1,
Congratulations You committed to buy the following item:

Motorola SLVR L7 L6 V8 Leather Case Pouch w/ Belt Clip

Sale price:	£0.99	
Quantity:	1	
Subtotal:	£0.99	
Postage	Seller's Standard Rate:	£3.99
Insurance:	£1.00 (optional)	

View item | Go to My eBay

Get Your Item

Pay Now

Click to confirm postage, get total price and arrange payment.

6 Listing items for sale

eBay is a great place to make some extra cash by selling unwanted or unused items that you have lying around the house or hidden in cupboards or garages. This chapter looks at finding items to sell, preparing them and then opening a seller's account on eBay.

Deciding what to sell

A lot of eBay users never sell anything on the site, while others use it as a method to raise some extra cash by selling unwanted items that are lying around in lofts and garages. Other people develop this even further and create businesses by selling items on eBay. Whichever group you fall into as a seller, your first task is to decide exactly what you are going to sell. Some areas to look at are:

- Items that have outgrown their usefulness. Areas to consider are old toys, children's clothes (as long as they are in good condition) and items relating to old hobbies.
- Items that have been replaced by newer products. Electrical goods and computer equipment are frequently being upgraded and there is usually a market for the older models.
- Unwanted presents. Most people have unwanted presents lying around in cupboards and drawers. These can easily be recycled by listing them on eBay.
- Family heirlooms. If you have a family heirloom that you dislike or do not want, it can be sold on eBay (but make sure you are not offending anyone by doing this).
- Collections. A lot of people collections from their childhood and which have since been left to gather dust. On eBay, collections can be one of the most profitable areas for sellers.

Before you start listing items to sell on eBay there are some points to consider:

- Make sure you do not mind parting with particular items and, if required, you have alternatives once you have sold them.
- The items really are yours to sell. Never try to sell anything that you do not own.
- Ask family members if you think anyone may be upset about anything you sell, such as heirlooms.

Beware

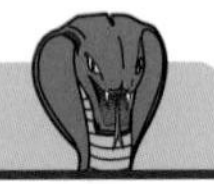

Just because you think a family heirloom is not worth keeping, this does not mean that all other family members will agree with you.

Hot tip

Collections are a very popular area on eBay. If you have a collection, or part of a collection, that you think may be valuable, take it to a professional valuation service or an auction house before you list it on eBay.

Sourcing items

The most successful eBay sellers are those who can source items from a variety of locations and keep on finding them. Some areas from which items for sale on eBay can be sourced are:

- Your own house. The closest source of possible items is within your own home. A quick look through cupboards and drawers can reveal a number of items that could be put up for auction. When garden sheds, garages and lofts are added there should be a significant, and varied, amount of saleable items.
- From friends and family. As well as selling the goods you find in your own house it is also worthwhile to ask around family and friends to see if they have anything they would like you to sell on their behalf. A lot of people have unwanted items but they are not interested in going to the trouble of actively selling them. If you offer to do this for them, you can then sell the items on eBay. However, make sure that you negotiate a cut of the price for yourself before you start selling.
- Car boot sales and jumble sales. Car boot sales and traditional jumble sales are an ideal way to obtain products that can then be resold on eBay. If you are going to do this, haggle to get a good price in order to maximise your profits on eBay.
- Antique shops. If you have a reasonable understanding of antiques then antique shops can be a valuable resource. However, it is vital to make sure that you know how much something is worth and that it is genuine. When selling antique items on eBay you will probably be asked a lot of questions by prospective buyers. If possible, try and get a certificate of authentication to provide reassurance for buyers on eBay.
- Live auctions. If you have an eye for a bargain then real auctions could be a good source of items that you can then sell on eBay. If you are participating in a live auction it is important that you find out how much

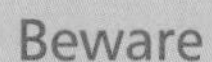

When selling for someone else make sure that they give you a proper description of the item and details about postage and packing. Remember, it will be your name that is against the auction.

you think an item is worth and set yourself a maximum price for bidding. Ideally, this should be relatively low so that you stand more of a chance of making a profit on eBay. If you are lucky there will be no other bidders and so you may get the item at a bargain price. If there are many bidders and the price is being pushed higher then it may be prudent to stop bidding.

Don't forget

When looking to buy items for resale, make sure that there is a market for them – specifically that there are already similar items for sale on eBay. Otherwise you could be left with unsold items.

- Small ads. Local newspapers usually carry small ads of items for sale and this can be another profitable area. Always go to see the items that are advertised and check out their condition. Also, always try to haggle the price down. In a lot of cases, people who are selling through the small ads just want to get rid of unwanted items and so they are generally happy to accept less than the advertised price.
- Discount shops. Almost all shops occasionally offer items at reduced prices, but some do this more frequently and with greater reductions. Cash and carry stores are ideal for picking up bargains: they frequently have significant discounts on a batch of items which can then be sold on eBay for a potential profit. Visit the shops regularly to find these types of bargains.

Don't forget

For more information about trading as a business see Chapter Eleven.

- End of season sales. Traditional shop sales are also good for picking up cheap items, although a bit of forward thinking is sometimes involved. The trick here is to buy seasonal items at the end of the season and then list them on eBay at the start of the season the following year. For instance, you could buy a batch of discounted swimwear in September and then list it on eBay in March or April the following year.
- Wholesalers. Wholesalers usually sell directly to the trade outlets, but if you can persuade one to sell to you then you could have a valuable asset as far as trading on eBay is concerned. If you are dealing with a wholesaler you will probably have to buy large quantities of items and pay for them promptly.

Preparing items for sale

Traditional retailers spend a lot of time and money ensuring that their goods look their best and are displayed as effectively as possible. In a similar way, it is important to prepare your items for sale on eBay as professionally as possible. This will not only help attract buyers but it will also cut down on the possibility of complaints from customers once a sale has been concluded. A little work put in before a sale can save a lot of time and hassle afterwards. Some points to consider when preparing items for sale:

- Make sure the product matches the description used on eBay. This is essential, as it is an area that can result in negative feedback from disgruntled buyers. If you know there is a defect in an item then say so in the description. The buyer will find out anyway once they receive the item, so it is better to be honest at the outset. When you are preparing the item see if you can spot any defects or faults that should be listed. Once you have prepared the item, note down a draft description and then check that it is an accurate reflection of the item.
- Ensure that items are clean and respectable. This is essential for items such as second-hand clothes, which have to be cleaned before they are listed on eBay. Even with other items it is important to make sure they are as clean as possible, even if this just means rubbing a damp cloth over something. With some items it is impossible to clean them completely, in which case this information should be included in the description.
- Make sure the item is working. For items such as electrical products or computer equipment, it is the seller's responsibility to make sure that the items are working properly when they are posted. Test items as thoroughly as possible; for example, if you are selling a television and video recorder in one auction, make sure they work individually and together.
- Take a photograph of the item, for inclusion on the site, after it has been prepared and is looking its best.

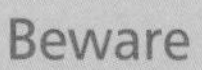
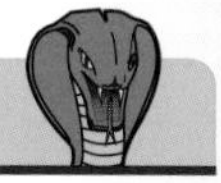

Beware

eBayers are generally very accommodating, but they can be highly critical of inaccurate or misleading descriptions of items.

Hot tip

If you are selling new items, try to make sure they are in their original packaging. This acts as an added guarantee of their condition.

Researching similar items

A bit of research is just as important for selling as for buying items. This gives you an idea of how much similar items have sold for, the condition they were in, the amount of postage charged and the number of bidders. Once you have this information it can be used to help determine the listing for your own items. The best way to research similar items from a seller's point of view is to look at completed auctions for these products. To do this:

Searching for items from completed auctions gives you an idea of the price they actually fetched, rather than a price mid-way through a particular auction.

1. Open the Advanced Search page and enter the item for sale. Check on the 'Completed listings only' box

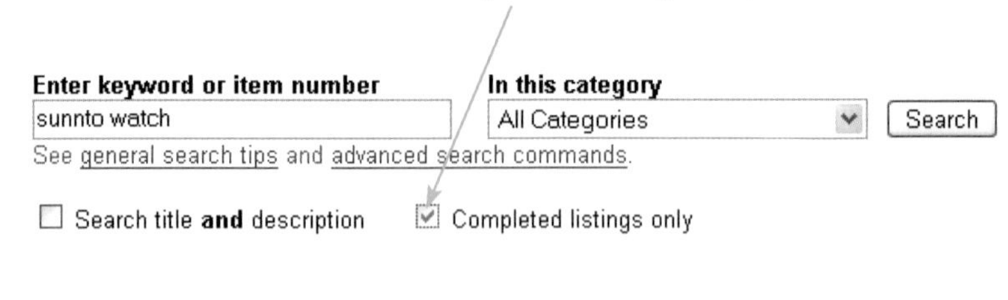

2. Click the Search button

3. A list of similar items is displayed in the search results. Click on one to view its details

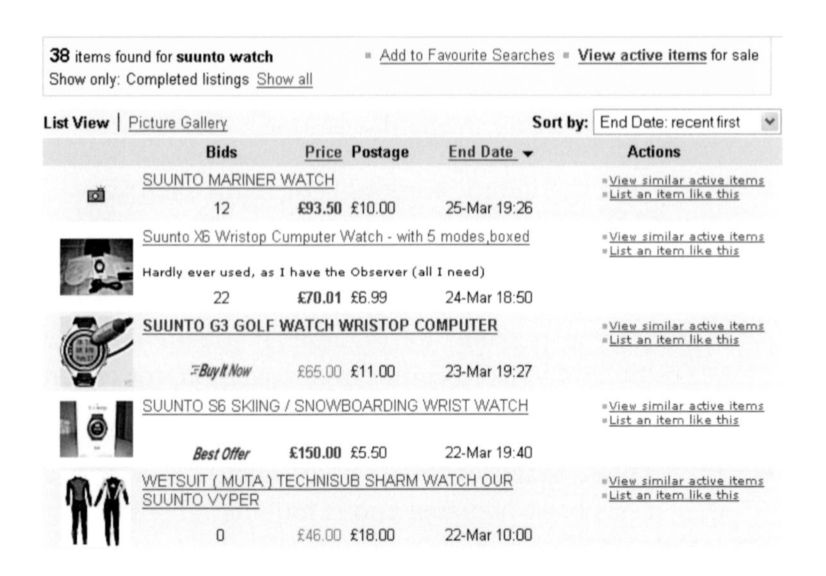

38 items found for suunto watch
Show only: Completed listings Show all
Add to Favourite Searches
View active items for sale
List View | Picture Gallery
Sort by: End Date: recent first
Bids Price Postage End Date Actions
SUUNTO MARINER WATCH
12 £93.50 £10.00 25-Mar 19:26
Suunto X6 Wristop Cumputer Watch - with 5 modes,boxed
Hardly ever used, as I have the Observer (all I need)
22 £70.01 £6.99 24-Mar 18:50
SUUNTO G3 GOLF WATCH WRISTOP COMPUTER
Buy It Now £65.00 £11.00 23-Mar 19:27
SUUNTO S6 SKIING / SNOWBOARDING WRIST WATCH
Best Offer £150.00 £5.50 22-Mar 19:40
WETSUIT (MUTA) TECHNISUB SHARM WATCH OUR SUUNTO VYPER
0 £46.00 £18.00 22-Mar 10:00
View similar active items
List an item like this

4. View the description and the photograph of the item to get an idea of its condition. This can then be compared with the price it fetched, which can be viewed here

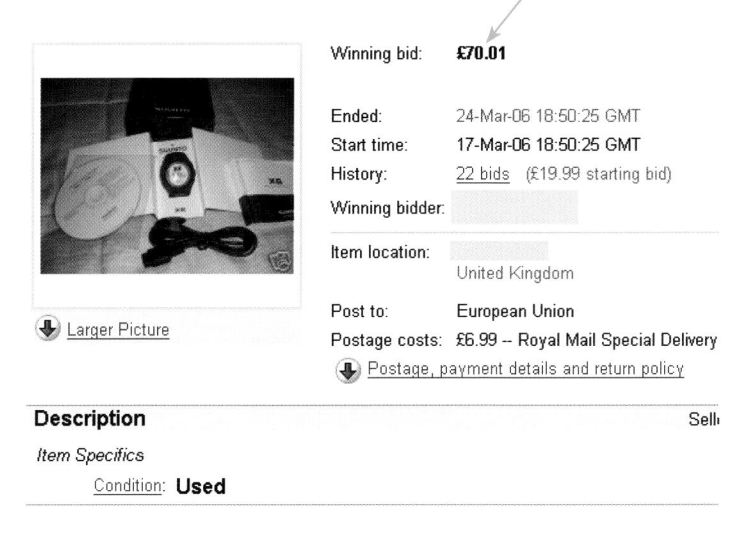

5. Check the postage and payment details. This gives you an idea of how much to charge for postage and the amount of flexibility that other sellers are offering

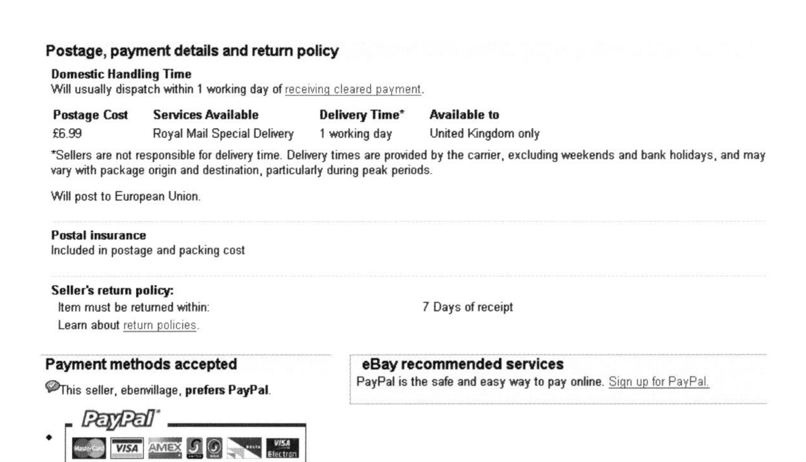

Postage, payment details and return policy
Domestic Handling Time
Will usually dispatch within 1 working day of receiving cleared payment.
Postage Cost
Services Available
Delivery Time*
Available to
£6.99
Royal Mail Special Delivery
1 working day
United Kingdom only
*Sellers are not responsible for delivery time. Delivery times are provided by the carrier, excluding weekends and bank holidays, and may vary with package origin and destination, particularly during peak periods.
Will post to European Union.
Postal insurance
Included in postage and packing cost
Seller's return policy:
Item must be returned within:
7 Days of receipt
Learn about return policies.
Payment methods accepted
This seller, ebenvillage, prefers PayPal.
PayPal
VISA
AMEX
eBay recommended services
PayPal is the safe and easy way to pay online. Sign up for PayPal.

The more payment options there are, the better it is for the buyer.

The seller's account page

The first step in becoming a seller on eBay is to set up a seller's account. This is obligatory as it provides eBay with the means to deduct the fees that are payable for every sale. There are three steps to start creating a seller's account:

1. Log in and click the Sell button on the home page main navigation bar

Don't forget

It is better to participate in a few auctions as a buyer before you start selling items. This will give you a feel for how eBay works and the types of things that happen in an auction.

2. The Selling introduction page is displayed

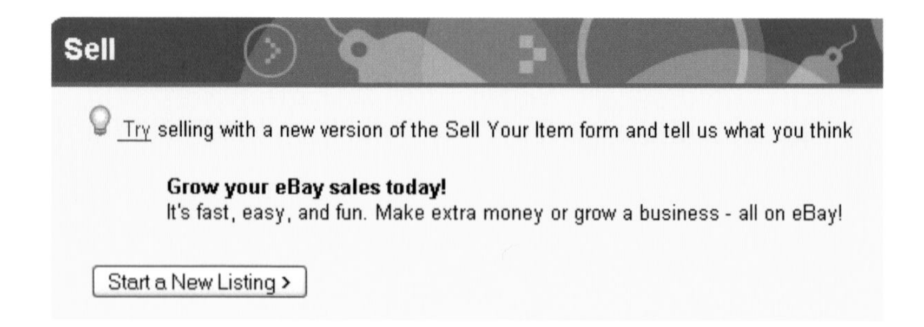

3. Click on the Start a New Listing button. This will take you to the sellers' account page

Start a New Listing >

Creating a seller's account

To set up a seller's account to start selling items:

1. Make sure you have your credit or debit card and bank account details to hand. These are used to verify your identification and also for the basis of the account itself

Seller's Account: Verify Information

1 Verify Information 2. Provide Identification 3. Select How to Pay Seller Fees

Please have the following ready, as you'll need both to sell on eBay UK:

- Credit or debit card
- Cheque book (bank account)

2. Enter your personal details. Although you have already entered these when you initially registered on eBay, they have to be entered again when you want to become a seller and create a seller's account

First, verify that your information below is correct.

First name: Nick

Last name: Vandome

Street

Town / City

County: -- Scotland --

Post code

Country: United Kingdom
Change country

Primary telephone ( )

Date of birth: --Day-- --Month-- Year

Continue >

3. Click Continue

Don't forget

eBay has strict rules about how they use personal details and they are committed to not passing them on to any third party that does not require them.

...cont'd

4. The next step is to verify your identification by matching your personal details held on eBay with those associated with your credit or debit card

Seller's Account: Provide Card Identification help

1. Verify Information (2) **Provide Identification** 3. Select How to Pay Seller Fees

Creating a seller's account is a **free** one-time process. However, you'll need to place a credit or debit card on file to become an eBay seller to help verify your identity. Your information is kept safe and private.

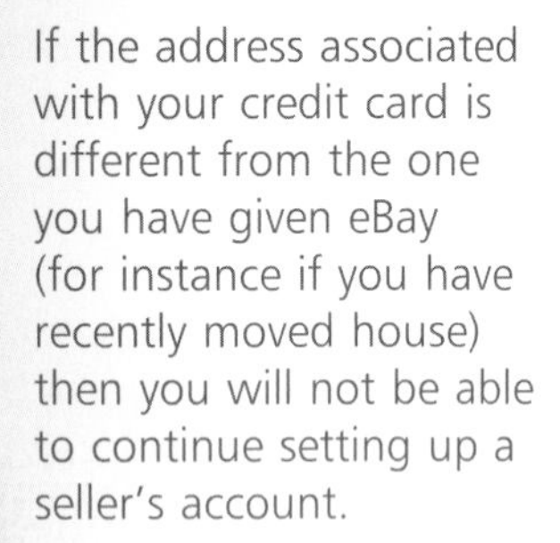

If the address associated with your credit card is different from the one you have given eBay (for instance if you have recently moved house) then you will not be able to continue setting up a seller's account.

5. Enter your credit or debit card details. At this point, this is just for verification purposes, although the information can be used later in the process to create the actual seller's account

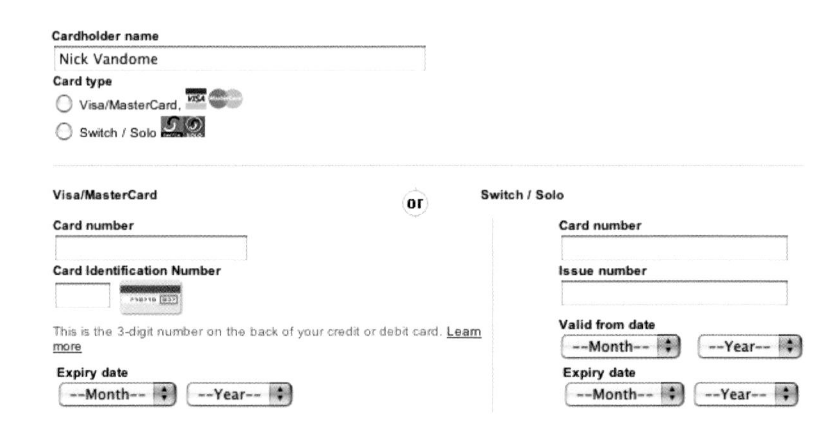

6. Click Continue

Your card will **not** be charged for creating a seller's account.

7. Enter the details of your bank account as further proof of your identify. This information can also be used for your seller's account, if you choose

Seller's Account: Provide Check Identification help

1. Verify Information 2 **Provide Identification** 3. Select How to Pay Seller Fees

Please place a bank account on file. Why? This assures eBay of your identity and increases overall site safety. eBay will **not** deduct funds from your account unless you authorise us to do so.

Account owner

Nick Vandome

Bank name

UK bank accounts only

Bank address

Bank postcode

8. Enter the sort code number and bank account number and click Continue

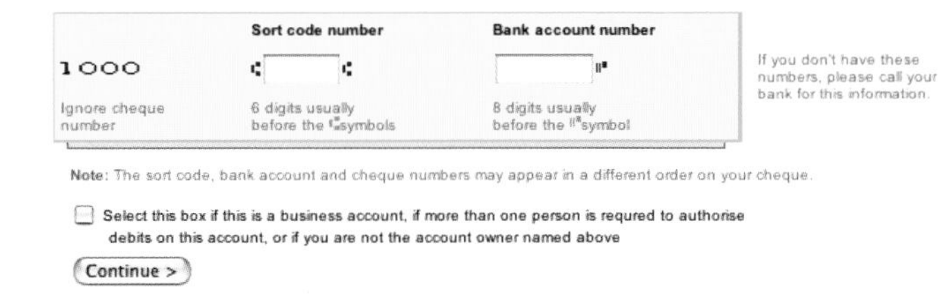

	Sort code number	Bank account number
1000		
Ignore cheque number	6 digits usually before the symbols	8 digits usually before the symbol

If you don't have these numbers, please call your bank for this information.

Note: The sort code, bank account and cheque numbers may appear in a different order on your cheque.

Select this box if this is a business account, if more than one person is requred to authorise debits on this account, or if you are not the account owner named above

Continue >

9. Select how you would like to pay the seller fees and click Continue. This will complete the setting up of your seller's account

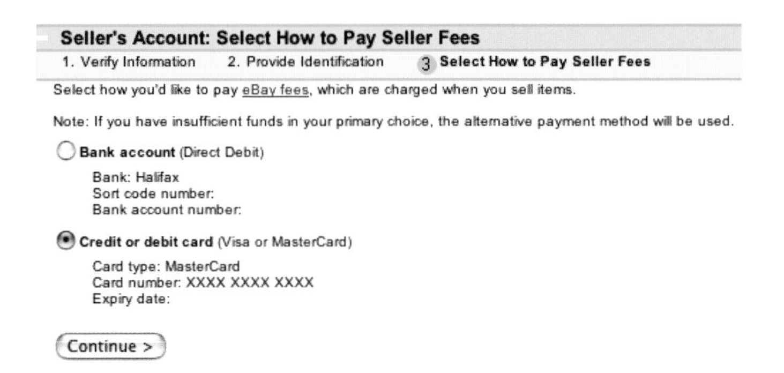

Seller's Account: Select How to Pay Seller Fees

1. Verify Information 2. Provide Identification 3 **Select How to Pay Seller Fees**

Select how you'd like to pay eBay fees, which are charged when you sell items.

Note: If you have insufficient funds in your primary choice, the alternative payment method will be used.

Bank account (Direct Debit)

Bank: Halifax
Sort code number:
Bank account number:

Credit or debit card (Visa or MasterCard)

Card type: MasterCard
Card number: XXXX XXXX XXXX
Expiry date:

Continue >

Don't forget

Seller fees are payable for all items that you sell on eBay. These fees are deducted each time you list an item and you can choose to pay them with either your bank account or credit/debit card.

Listing items to sell

Once you have created a seller's account you can then start listing items to sell. The process is begun in the same way as shown on page 96 for initially setting up a seller's account. Then an item can be listed for sale as follows:

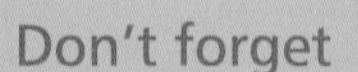

If you do not want to sell your item in an auction, check on the Sell at a Fixed Price button. This will make your listing a Buy It Now sale only.

1. Select this button to list the item for sale in an auction

Sell Your Item: Choose a Selling Format

To begin, select a format and click the **Continue** button. Please make sure your item is allowed.

- **Sell at online Auction**
  Allows bidding on your item(s). You may also add the Buy It Now option. Learn more.
- **Sell at a Fixed Price**
  Sell your items instantly at a Fixed Price using Buy it Now. Note: These items are now available on all eBay sites. Learn more.

Want someone else to sell for you? Find a Trading Assistant.

Continue >

2. Click Continue

Continue >

3. Click here to select a main category for the item. This is the category under which it will appear in the eBay categories listings

Main category

Click to select	
Antiques & Art	Furniture, metalware, oriental antiques and antiquities.
Baby	Baby Clothes, Baby Toys, Nursery Equipment & Furniture
Books, Comics & Magazines	Fiction and non-fiction books, textbooks, comics and magazines
Business, Office & Industrial	Building materials & supplies, power/industrial tools, restaurant equipment, business for sale & more
Cars, Parts & Vehicles	Cars, Parts, Motorcycles, Boats & Other Vehicles
Clothes, Shoes & Accessories	New and used branded clothes, shoes & accessories.
Coins	Coins, banknotes, bullion, historical medals and tokens.
Collectables	Animals to animation, militaria to memorabilia, postcards to trading cards
Computing	Desktops, Laptops, PDAs, Components, peripherals, software, servers & much more
Consumer Electronics	Audio systems, portable audio, DJ Equipment, TV & Home video, & much more
Crafts	Art equipment, Knitting, Sewing & Cardmaking
Dolls & Bears	Dolls, dolls' houses, dolls' house contents and teddy bears
DVD, Film & TV	DVDs, videos, film and TV memorabilia
Health & Beauty	Fragrances, Dietary Products, Skincare & Make-up
Home & Garden	Home furnishings, household equipment, DIY & garden items
Jewellery & Watches	Fine jewellery, antique, artisan, and supplies

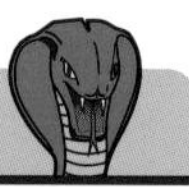

Try not to use the Everything Else category, as it is a bit vague when people are searching for items. There are enough eBay categories that you should be able to find one to match the item you are selling.

4. If the item is not in any of the main categories, select the Everything Else option at the bottom of the page

Everything Else — Weird, wonderful, and unusual

5. Alternatively, enter a description of the item here and click Search

Enter item keywords to find a category

suunto [Search] Tips

For example, "gold bracelet" not "jewelery"

6. A category listing is pre-selected, based on the information entered in the search box

Category	
○ Sporting Goods > SCUBA & Snorkelling > Dive Computers	(48%)
○ Sporting Goods > SCUBA & Snorkelling > Other SCUBA	(10%)
○ Sporting Goods > SCUBA & Snorkelling > Gauges	(9%)
○ Jewellery & Watches > Watches > Wristwatches > Men's > Other Men's Watches	(5%)
◉ Jewellery & Watches > Watches > Wristwatches > Other Wristwatches	(5%)
○ Sporting Goods > SCUBA & Snorkelling > Regulators	(5%)
○ Jewellery & Watches > Watches > Other Watches	(3%)
○ Sporting Goods > Hiking > Equipment > Compasses	(3%)
○ Sporting Goods > Exercise & Fitness > Monitors/Pedometers/Computers	(2%)
○ Sporting Goods > SCUBA & Snorkelling > Buoyancy Compensators	(2%)

[Cancel] [Sell In This Category]

Tip: Add a second category to increase your item's exposure. You can do this at the bottom of the main category page.

7. Click Sell In This Category

[Sell In This Category]

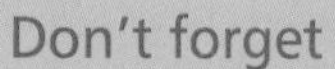

If you do not search for a category using the Search box, you will be able to select a sub-category for your item after you have selected a main category.

Writing a description

One of the most important elements in a listing to sell an item is the description. This tells prospective buyers about the item, the condition it is in and any additional information you want to disclose. To add a description of an item:

The item title appears in the search results page.

1. Enter the title here. This should state exactly what the item is

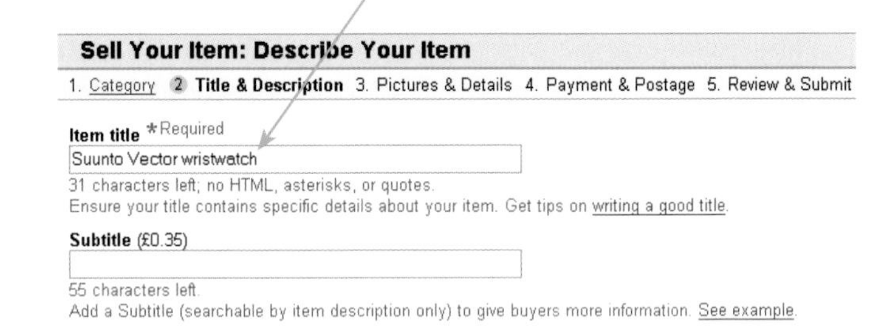

Sell Your Item: Describe Your Item
1. Category 2 Title & Description 3. Pictures & Details 4. Payment & Postage 5. Review & Submit
Item title *Required
Suunto Vector wristwatch
31 characters left; no HTML, asterisks, or quotes.
Ensure your title contains specific details about your item. Get tips on writing a good title.
Subtitle (£0.35)
55 characters left.
Add a Subtitle (searchable by item description only) to give buyers more information. See example.

Don't forget

Be as honest as possible when writing descriptions. In the long-term, you will gain little or no advantage by not describing items accurately.

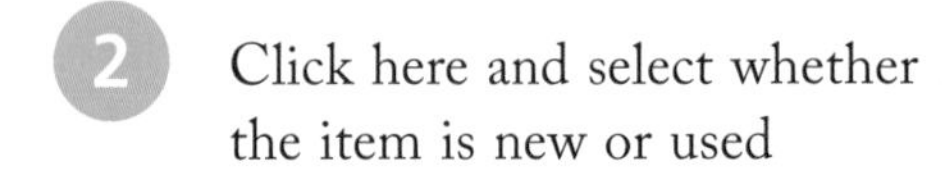

2. Click here and select whether the item is new or used

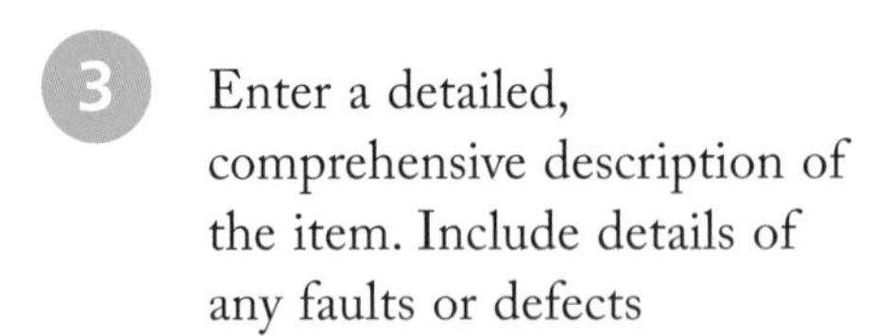

3. Enter a detailed, comprehensive description of the item. Include details of any faults or defects

4. Click Continue to move to the next step

Item description *

Describe your items features, benefits, and condition. Be sure to include in y dimensions or size. You may also want to include notable markings or signa Wristwatches.

Suunto Vector wristwatch with compass, altimeter and barometer. Very good condition and includes full instruction manual.

Enter <p> to start a new paragraph. Get more HTML tips.

Spell check | Preview description

Note: After you click **Continue**, you may be asked to download a small file. page does not appear, change your picture selection method.

Pictures, price and duration

The next stage of the process is to add images to accompany the description of the item, determine the starting price for the item and set the duration of the auction. To do this:

1. Enter the price in the Starting price box

Pricing and duration

Price your item competitively to increase your chance of a suc

NEW! Get ideas about pricing by searching completed items...

Starting price *Required

£ £5.00

A lower starting price can encourage more bids.

2. Click here and select the duration for the auction. The minimum is 1 day and maximum is 10 days

3. Select this option to have the auction start as soon as the item is submitted for listing

4. Select this option to select a specific time for the start of the auction. This gives you more control over when an auction starts and, more importantly, when it finishes

Don't forget

The starting price is the price at which the auction for an item will begin. This is the minimum price that a buyer has to initially bid for an item.

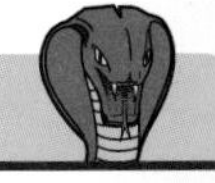

Beware

Be careful about setting a starting price that is artificially low in the hope of attracting a lot of bidders. If you only get one bid at the starting price then this is the amount you will have to sell the item for (unless you also set a reserve, as explained on page 114).

Hot tip

Try to ensure that your auction goes over at least one weekend, when there are likely to be more people with time to look on eBay.

Hot tip

If possible, make sure the ending time for an auction is when you will be able to access eBay. This way you will be able to contact the winning bidder immediately.

...cont'd

To list multiple items in a single auction, also known as a Dutch auction, you must have your direct debit details on file, or have an overall feedback rating of 20 or higher, or if you have a PayPal account and an overall feedback rating of 10 or higher.

5. Scroll down the page and select the quantity for the item you have for sale. The default is one, but if you have several identical items to sell, click the Multiple Item link (if you are eligible)

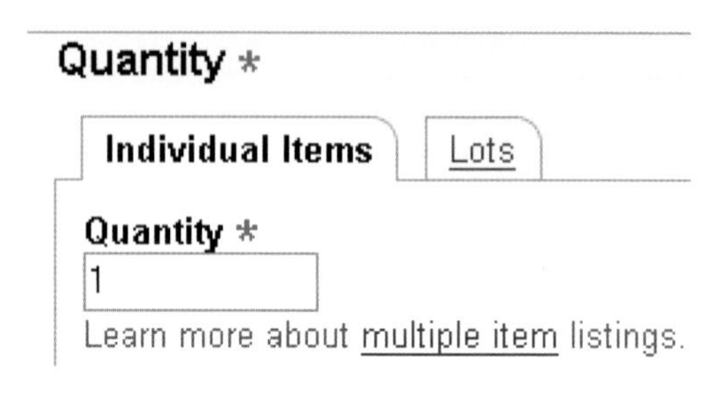

6. By default, the location of the item is that given at the time of registration. Click Change to send it from a different location

Item location

Postcode: Not specified
Location display: Perth, United Kingdom
Change

7. Click here to select an image to accompany the item's description

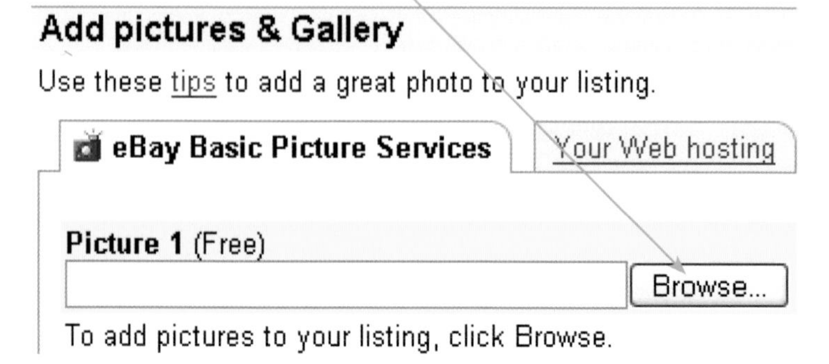

8. Browse your hard drive, select an image and click Open (Windows) or Choose (Mac)

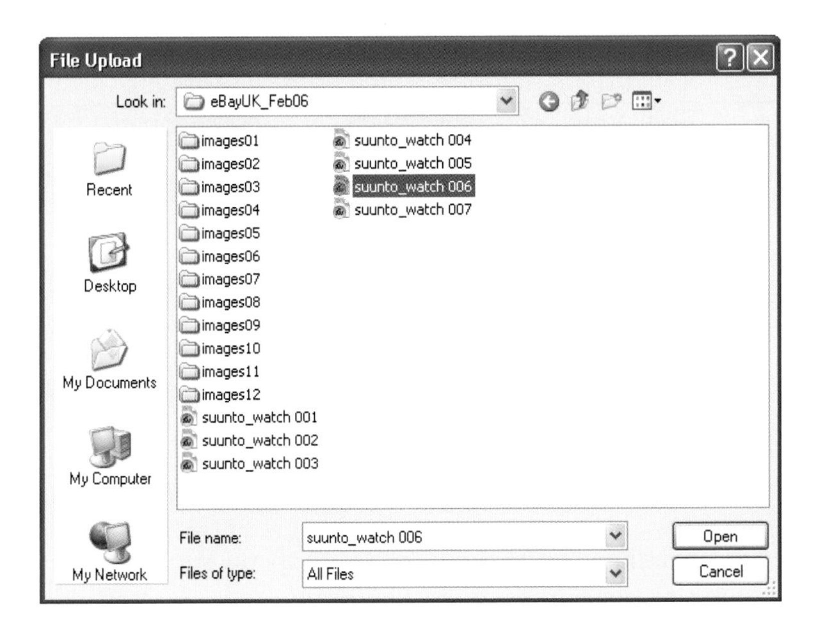

9. The selected image is noted here. This will appear next to the description of the item when the listing is submitted

Hot tip

You can add an image from an existing website of your own, if you have one. To do this, click on the Your Own Web Hosting link and enter the full URL (Web address). Make sure this is for the image (with a .jpg, .gif, or .png extension) rather than a Web page (with an .htm extension).

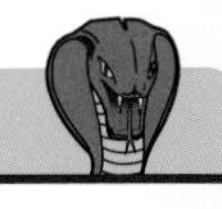

Beware

It is strictly prohibited to link to images on other people's websites, unless you get their permission.

Enhanced options

eBay offers a number of enhanced options to make a listing stand out more from the crowd. To use the enhanced options:

Hot tip

It is always worth selecting the Gallery option because buyers like to see pictures of items as soon as the search results page appears.

1. Tick these boxes to have your listing displayed in the Gallery site, where the images are displayed more prominently

Gallery options
Applies to first picture
☑ Gallery (£0.15)
Add a small version of your first picture to Search and Listings. See example.
☐ Gallery Featured (£15.95)
Add a small version of your first picture to Search and Listings and showcase
See example.
☐ Remember Gallery selection next time I sell.

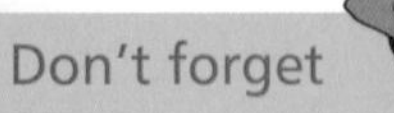

Don't forget

There is a Supersize option for displaying images at a larger size.

2. Select one of these options to specify the size at which your pictures are displayed

Picture options
Applies to all pictures.
◉ Standard
Standard pictures will appear within a 400 x 400 pixel area.
☐ Supersize Pictures (£0.60)
Extra large pictures will appear within a 500 x 500 pixel
☐ NEW! Picture Show (£0.15)
Multiple pictures will appear in a slideshow player at the
○ Picture Pack (£0.90 for up to 6 pictures or £1.35 for 7 to 12
Get Gallery, Supersize, Picture Show and additional pictur

Don't forget

The more enhanced options that are selected, the greater the overall cost of a listing.

3. Click here to select a theme for how the listing is displayed

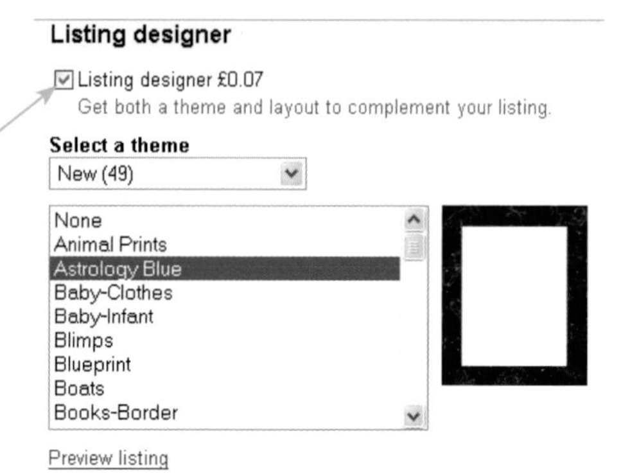

4. Click here to have your listing appear in bold

Add Gallery above in pictures
Bold (£0.75)
Attract the attention of buyers
Highlight (£2.50)
Make your listing stand out wi

5. Click here to have your listing appear with a coloured band behind it to make it stand out more

Don't forget

Emphasising items is looked at in more detail in Chapter Seven.

6. If you want a page counter with your listing, select one of the options here. This will show how many visitors you have had to your listing once the auction starts

Page counter
No counter
1234 Andale Style
1234 Green LED

Hot tip

Including a page counter is always worthwhile as it is free and it gives you an idea about the level of interest your auction is generating.

7. Click Continue to move to the next step of the listing process, or Back to edit some details

< Back   Continue >

Payment and postage

Any buyer should pay careful attention to the amount of postage and the methods of delivery. Therefore it is important for the seller to give some thought to these and to try to make the entire listing as attractive as possible to any prospective buyers. To enter the payment and postage details:

1. Tick this box if you want buyers to have the option of using PayPal to pay for an item

Sell Your Item: Enter Payment & Postage

1. Category 2. Title & Description 3. Pictures & Details 4 P

Title
Suunto Vector wristwatch

Payment methods

Choose the payment methods you'll accept from buyers.

PayPal enables you to accept credit card payments in multiple

☑ PayPal payment notification will go to: nickvandome@mac
No account needed. Fees may apply

2. Tick the other payment options which you are willing to accept

Other payment methods

☑ Postal Order / Banker's Draft

☑ Personal Cheque

☐ Other / See Item Description

☐ Credit Cards

☐ Escrow
eBay's approved provider is escrow.com.

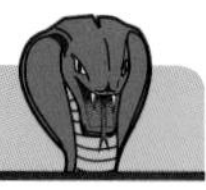

Beware

Don't make postage prohibitively expensive or try to subsidise a low starting price for an item by increasing the postage cost. This practice is not allowed by eBay.

Don't forget

PayPal is a means of transferring money electronically between buyers and sellers. It is the most commonly used method of payment on eBay.

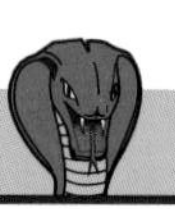

Beware

Never accept cash as a means of payment, even if it is suggested by the buyer as their only method of payment.

Don't forget

For a more detailed look at payment options see Chapter Ten.

3. Select an option for where you are willing to send the item. Click here if you only want to ship the item to a location in the UK

Post-to locations *Required

Will post to United Kingdom and the following (check all that apply):
Reach more buyers - learn more about posting internationally.

- Worldwide
- N. and S. America
  - United States
  - Canada
- Europe
  - European Union
  - Ireland
  - France
  - Germany
- Asia
  - Australia
  - Japan

Will not post - local pickup only
Specify pickup arrangements in the Payment Instructions box below.

4. Click here for postage options

Postage *

Specify a flat cost for each postal service you offer.

Domestic Postage (offer up to 3 services)

Select a postage service £
Select a postage service £
Select a postage service £

5. Select a method of postage

Select a postage service
Seller's Standard Rate
Royal Mail 1st Class Standard (1 to 2 working days)
Royal Mail 2nd Class Standard (1 to 3 working days)
Royal Mail 1st Class Recorded (1 to 2 working days)
Royal Mail 2nd Class Recorded (1 to 3 working days)
Royal Mail Special Delivery (1 working day)
Royal Mail Standard Parcels (3 to 5 working days)
Parcelforce 24 (1 working day)
Parcelforce 48 (2 working days)
Other Courier
Collection in Person

Beware

Even if you state that you are only willing to post items to one location you may still have people from around the world bidding on your auction.

Don't forget

Try to have more than one option for the method of postage. This will allow the buyer to select a cheaper option if they want.

...cont'd

Hot tip

If you are in any doubt about the amount of postage to charge, take the item to the post office, get it weighed and find out exactly how much it is for different postal services.

Don't forget

Some sellers add a small amount to the actual cost of postage to cover their time spent packaging and posting it.

Don't forget

If you include postal insurance this will be an additional charge to cover insurance for the item if it is lost or damaged while in transit. This is most frequently used when sending items abroad.

Hot tip

Having a transparent and robust return policy is a good way of making buyers feel confident about doing business with you.

6. Enter an amount for postage and packing

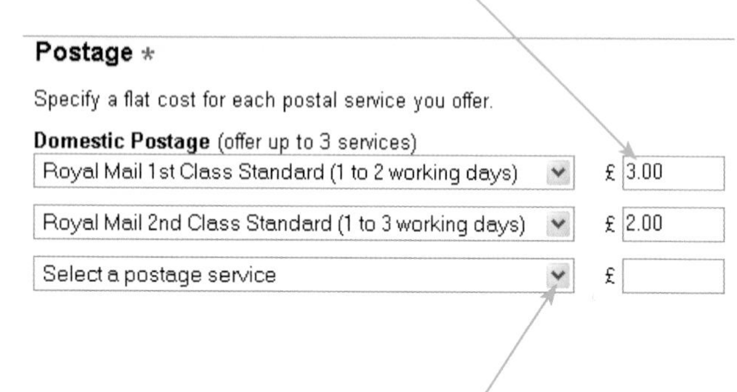

7. Click here to select another postal option. This can be set in the same way as the first ones

8. Click here if you want to offer postal insurance

Postal Insurance
Not offered £
Postage Discount
Make this item eligible for a postage discount based on my preferences.
Discount will be applied only if the buyer purchases other items from you.

9. Enter specific payment instructions and any details you want to note about a return policy for the item

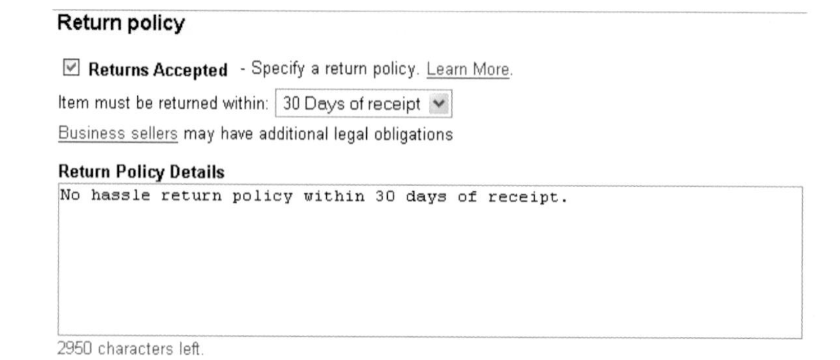

Reviewing a listing

At the end of the listing process it is possible to review all of the details you have entered for the item you are selling:

1. Review the title, description and images used. Try to look at these with fresh eyes and imagine how prospective buyers will view these details

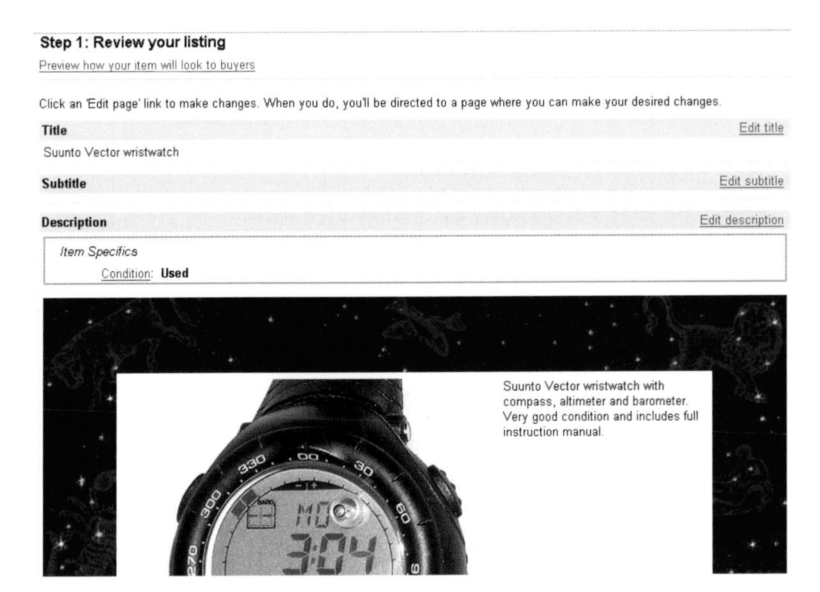

Step 1: Review your listing

Preview how your item will look to buyers

Click an 'Edit page' link to make changes. When you do, you'll be directed to a page where you can make your desired changes.

Title Edit title

Suunto Vector wristwatch

Subtitle Edit subtitle

Description Edit description

Item Specifics

Condition: **Used**

Suunto Vector wristwatch with compass, altimeter and barometer. Very good condition and includes full instruction manual.

2. Scroll down the page and review the fees you will be charged once the listing is submitted

Step 2: Review the fees and submit your listing

Note: Fees charged for scheduled listings are those applicable at Your Item form.

Listing fees (Incl. VAT)	
Insertion fee:	£ 0.35
Scheduled start time:	0.06
Gallery:	0.15
Listing Designer:	0.07
Total added listing fees: *	**£ 0.63**

Don't forget

The Insertion Fee is based on the starting price for an item. It is non-refundable even if the item does not sell. For more information on fees, see Chapter Ten.

...cont'd

3. If you are happy with the details of the item, click Submit Listing. If you want to make changes click Back

< Back   Submit Listing

4. Once you have submitted your listing you will see the following confirmation message

5. The item will be displayed in the Items I'm Selling section on your My eBay page

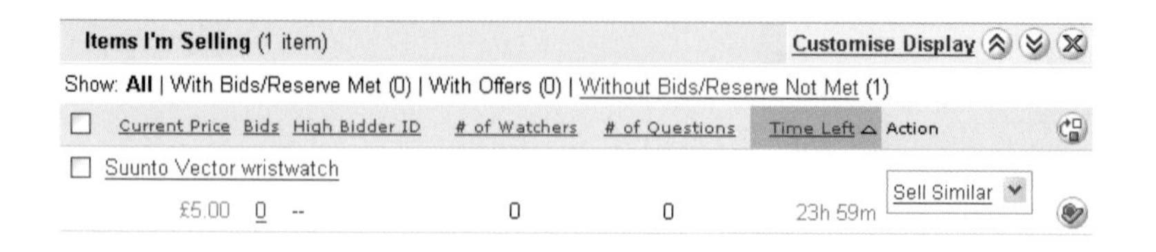

7 Advanced listing

This chapter looks at how to create a listing to give an item the best possible chance of selling and obtaining a good price.

To reserve or not to reserve

One of the decisions that you will have to make when listing an item for sale is whether to include a reserve price. This is a minimum price at which you are prepared to sell the item: once the auction is completed you do not have to sell the item if the reserve price has not been met. Buyers can see that there is a reserve price attached to an item, but not how much it is. The minimum value for a reserve price is £50.00. The advantage of a reserve price is:

- You are guaranteed a minimum amount for an item; if this is not met, you do not have to sell the item.

The disadvantage of a reserve price is:

- It can put off buyers; since they cannot see the amount of the reserve price, they are unsure of how much they have to bid just to meet the reserve, never mind to win the auction.

Don't forget

If you set a reserve price for an item, this will be the figure upon which the insertion fee will be calculated, not the starting price.

When a reserve is set for an item, a starting price for the auction is also listed. This is usually significantly lower than the reserve price: the idea is that bidders will be attracted by the low starting price and the bidding will go above the reserve. In practice it is usually better to have no reserve and a higher starting price. Until the reserve price is met, there is a note underneath the item stating 'Reserve Not Met'. The reserve price is entered during the listing process.

To add a reserve price to an item for sale:

1. Enter a starting price for the auction

Starting price *Required
£ 5.00
A lower starting price can encourage more bids.

2. Enter the reserve price: the minimum amount which you are prepared to accept for the item

Reserve price (fee varies)
£ 50.00

Capturing images

No matter how good a textual description of an item is, a good picture can make the difference between a successful, profitable sale and an auction with no bidders at all. There is a lot of competition on eBay so everything that you can do to give yourself an advantage is important.

When capturing images for use in an auction, there are some points that you should follow to get the best possible initial image:

- Get as close to the object as possible. Buyers want to see the object itself, not the background.
- Make sure the background is as neutral as possible and does not detract from the item itself.
- Make sure that images are in focus. If you find an image is out of focus, take another one. A bit of effort at this stage could pay dividends later.
- Capture the item at different angles. You can include up to a dozen images with a listing (at a cost of 12p each after the first one) so make the most of this and include shots from the front, back, top and side, if applicable. You may not use all of the images but it is good to have a choice.
- Use natural light if possible. The best lighting for taking photographs occurs either first thing in the morning or just before dusk. This produces a natural, soft light that is ideal for photography. If you cannot manage this, then still try and take images in daylight. If you have to take them inside then use a flash, but make sure there is no glare reflected off the subject. This can be a particular problem with glass objects.
- If you are using a digital camera, use a low resolution setting, such as 640 x 480. This will not affect the quality of the image when it is used on eBay, but it will create a smaller file size, which will speed up the downloading time. Similarly, if you are scanning a hard-copy image, use a low resolution.

Hot tip

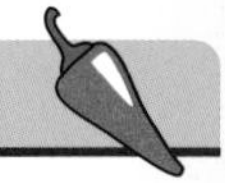

If you are not a confident photographer, enlist the help of a family member or a friend who is more proficient with a camera, and ask them to take some pictures of your item.

Hot tip

Use a tripod if possible, to make sure the picture is composed properly and that the image is in focus and does not suffer from camera shake.

Sample images

1. This is a poor image because the subject is too small, is slightly obscured and the background is too dark (there is not enough contrast between the subject and the background)

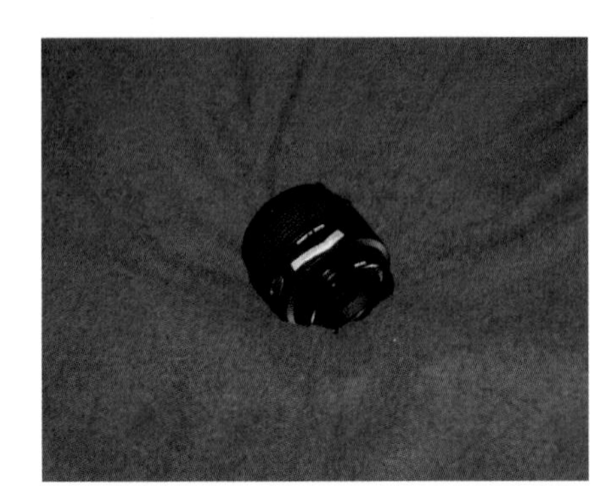

2. This is a better composition because the subject takes up more of the image and the back is more neutral so it does not distract from the subject. However, it is slightly out of focus which spoils the overall effect

Hot tip

Most images can be improved by some simple digital image editing. See the following pages for details.

3. This is the type of image that should be used to accompany a listing. It is sharp and crisp and the subject is prominently displayed

Editing images

Even if you capture an excellent initial image, it will probably benefit from some digital editing in an image editing program such as Photoshop, Photoshop Elements, Paint Shop Pro or Ulead Express. Even some basic techniques can produce dramatic results.

Editing Levels

The levels in a digital image determine the distribution of the lightest and darkest pixels in the image, spread across a graph known as a histogram. If there are not enough pixels at the extremes of white and black, the image can appear either over- or under-exposed. If this happens, the Levels command in a digital image editing program can be used to redistribute the pixels so that there are some at both extremes of the graph. This has the result of improving the overall tone of the image. In some programs this can be done with an Auto Levels command, but it can also be done manually. To do this:

1. Open an image and access the Levels dialog window

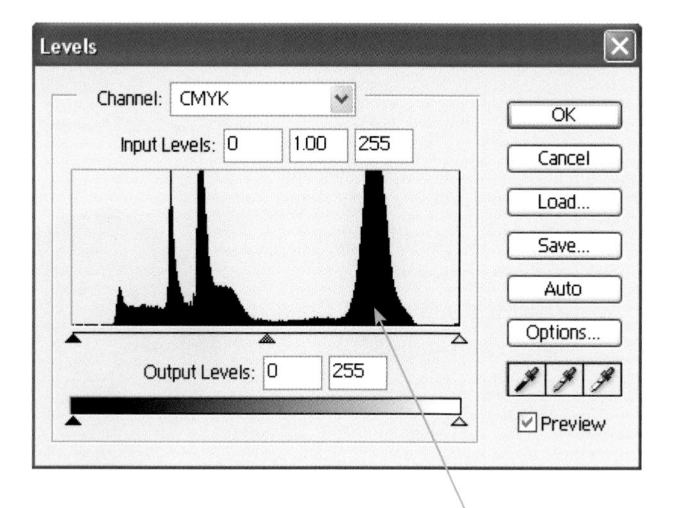

2. The tonal range of pixels in the image is shown here. If there are gaps at either end it means the image is either too dark or too light at the extremes of this graph

Don't forget

The example screen shots in this chapter use Photoshop Elements, but similar features exist in other image editing programs.

Don't forget

A pixel (which is a contraction of 'picture element') is a coloured dot in a digital image. Most images are made up of thousands, or millions, of pixels.

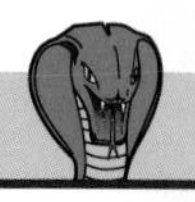

Beware

Never use image editing software to change the physical appearance of an item you are selling. This is dishonest and could result in problems once the sale has been completed. Only use image editing software to enhance what is already in the picture.

Don't forget

For a detailed look at digital photography and digital imaging, have a look at 'Digital Photography in Easy Steps'.

...cont'd

3. Drag these sliders to improve the tonal range of the image

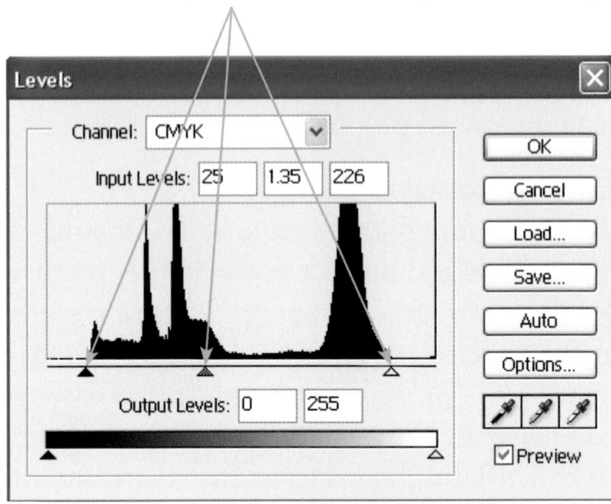

Hot tip

Most image editing programs have an Auto Levels option which will adjust the level automatically. This is worth trying, but, generally, the results are not as accurate as doing it manually.

4. An image before Levels correction has been applied. It generally looks a bit 'flat' with a limited tonal range

5. Once Levels correction has been applied, the image has a much better tonal range and the main subject stands out more

Brightness and Contrast

Another simple improvement to an image can be made by adjusting the brightness and contrast. To do this:

1. Open an image that looks either too dull or too bright

2. Access the Brightness/ Contrast dialog window and drag the sliders

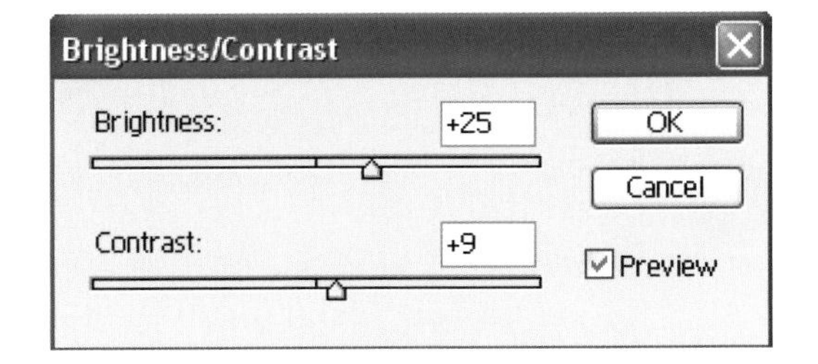

3. The final image now looks much brighter and sharper

Hot tip

Tick the Preview box to see the changes take effect as they are made in the dialog box.

...cont'd

Sharpening

Sharpening is a technique that improves the clarity of an image. Although it can be used to try to make an out of focus image a bit clearer, it is most effective on a standard digital image that appears in reasonable focus. A lot of digital cameras apply some sharpening to images as they are captured and if this is also done in an image editing program it can give an already acceptable image added clarity. To apply sharpening:

Don't forget 

Sharpening works by increasing the contrast between the edges of adjoining pixels in an image. This creates a greater definition between pixels, resulting in a sharper image.

1. Open an image that is acceptable but either slightly blurry or marginally out of focus

Don't forget 

In addition to the standard sharpening, there are additional options for applying greater degrees of sharpening. These are Sharpen More, Sharpen Edges and Unsharp Mask.

2. Select the Sharpening option (usually found under the Filter menu in image editing programs)

3. The image appears clearer and more sharply defined

Optimising images

Optimisation is a term given to the technique of making image files as small as possible, while still retaining a suitable quality. This is important for images that are used on the Web because the larger the file size, the longer it takes for them to download. Also, most consumer digital cameras are now capable of taking large images, in terms of both file size and physical dimensions. This is excellent for printing, but unnecessary for the Web since the files will be viewed at sizes and resolutions that are far smaller than in the original image. Although eBay does not stipulate that images have to be under a certain file size when they are sent for inclusion with a listing, it is a good idea to try to make them as small as possible. This is done by changing the physical size of the image by removing unwanted pixels. To do this:

1. Open the Image Size dialog window. The current dimensions are shown here. Tick the Constrain Proportions and Resample Image boxes

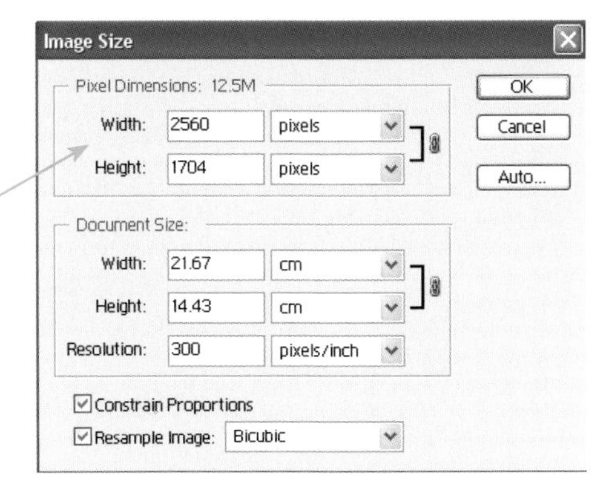

2. Enter a new figure for the Width, and the height will be adjusted proportionally. This will also reduce the file size of the image. Click OK to apply the changes

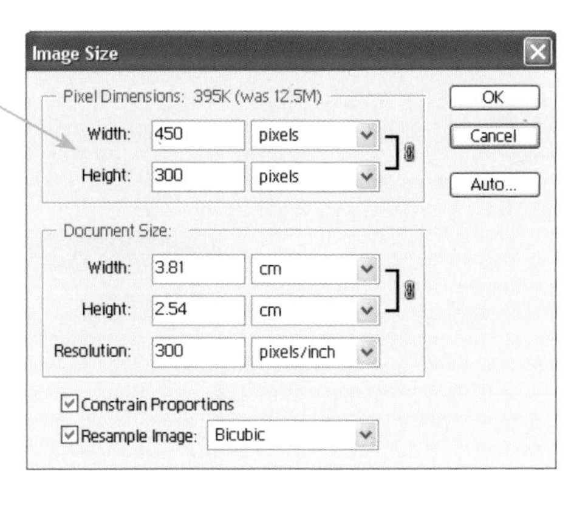

Don't forget

Tick the Resample Image box to enable the size of the image to be changed. Tick the Constrain Proportions box to change the size of the image proportionally.

Make sure the image is big enough for any picture options that have been selected in eBay: a standard picture appears in an area of 400 x 300 pixels, while a supersize picture appears in an area of 500 x 375 pixels. Do not reduce the image size beyond these dimensions.

Using HTML

HyperText Markup Language (HTML) is a markup code which is used to format the vast majority of the pages on the Web. It is not a full-blown programming language but rather a way of displaying information. It uses tags to instruct Web browsers how certain items on the page should be displayed. These tags are contained within angled brackets and there are usually, but not always, an opening and a closing tag. So if you want to produce bold text on a Web page you could use:

<b>Welcome to my Web page</b>

When adding descriptions of items in an eBay listing it is possible to use HTML code to make the text look more interesting and stand out more. There are a number of simple HTML tags that can be added within the description box of an eBay listing:

If you are familiar with HTML you can include your own markup for an item description. This will show up once the item is listed.

Bold

1. Add a <b></b> opening and closing tag to create bold

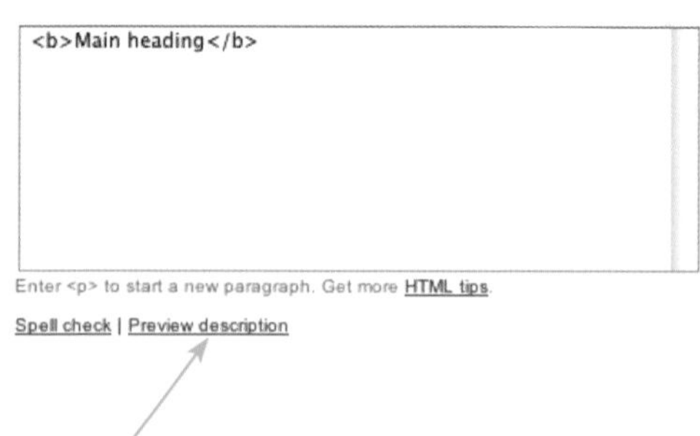

2. Click Preview Description to see how the effect will look

For more information about HTML, see 'HTML in Easy Steps'.

Italics

1. Add an <i></i> opening and closing tag to create italics

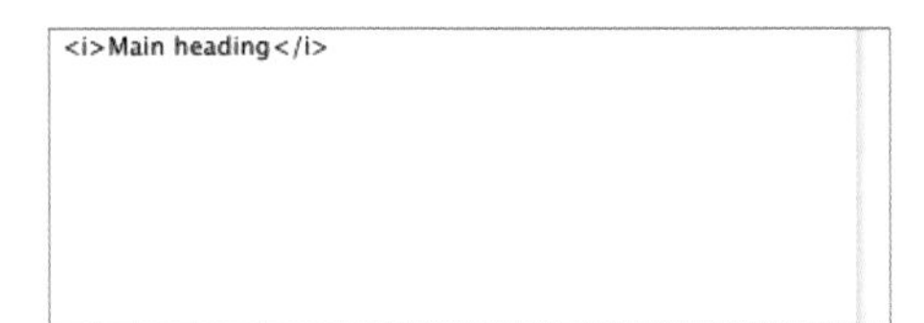

2. Click Preview Description to see how the effect will look

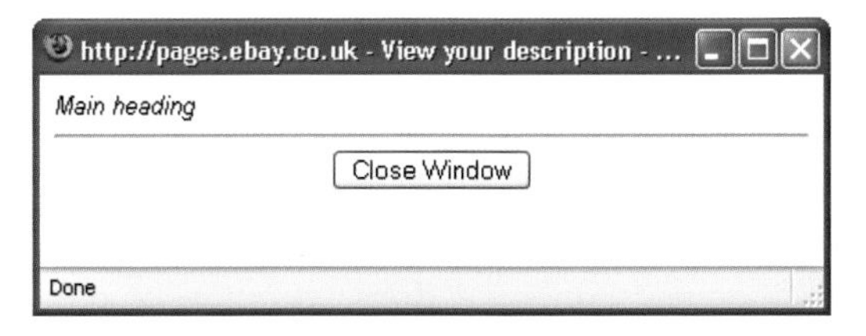

Main heading

1. Add an <h1> </h1> opening and closing tag to create a main heading

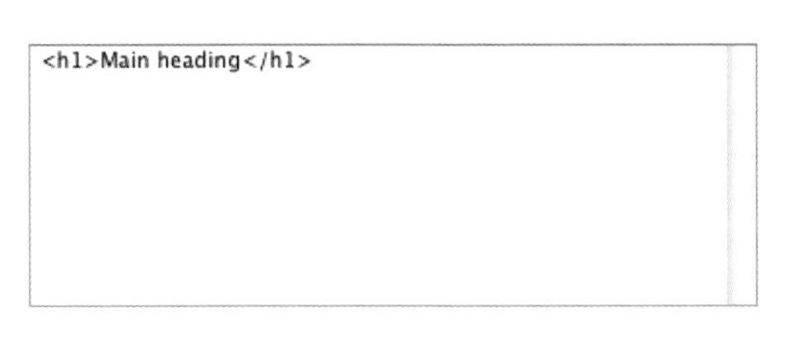

2. Click Preview Description to see how the effect will look

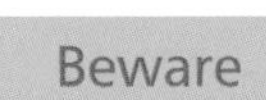

Beware

Don't use underlining in HTML as this could be mistaken for a hyperlink once it is displayed on an eBay Web page.

...cont'd

Sub heading

1. Add an <h2></h2> opening and closing tag to create a sub heading

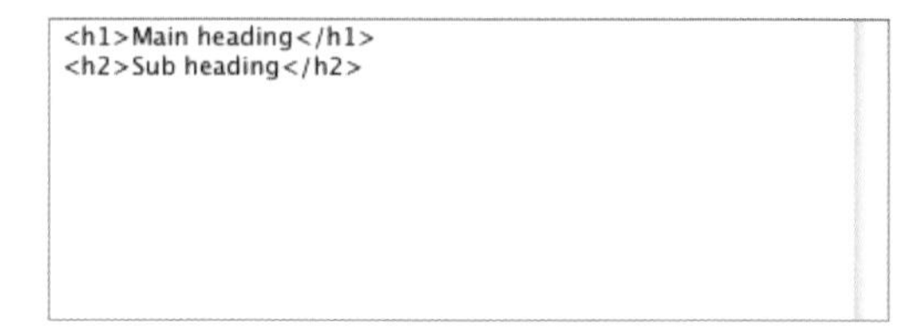

```
<h1>Main heading</h1>
<h2>Sub heading</h2>
```

2. Click Preview Description to see how the effect will look

Don't forget

There are six levels of headings within HTML.

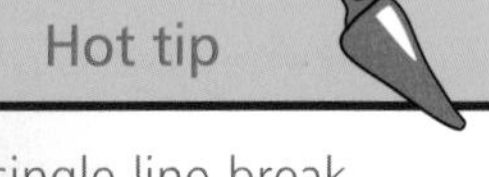

Hot tip

A single line break can be created by including the <br> tag. This does not have a corresponding closing tag.

New paragraph

1. Add a <p></p> opening and closing tag to create a new paragraph

```
<h1>Main heading</h1>
<h2>Sub heading</h2>
<p>Main content goes here</p>
```

2. Click Preview Description to see how the effect will look

Font colour

1. Add a <font color> </font> opening and closing tag to create a text colour

```
<h1>Main heading</h1>
<h2>Sub heading</h2>
<p><font color=green>Main content goes here</p></font>
```

2. Click Preview Description to see how the effect will look

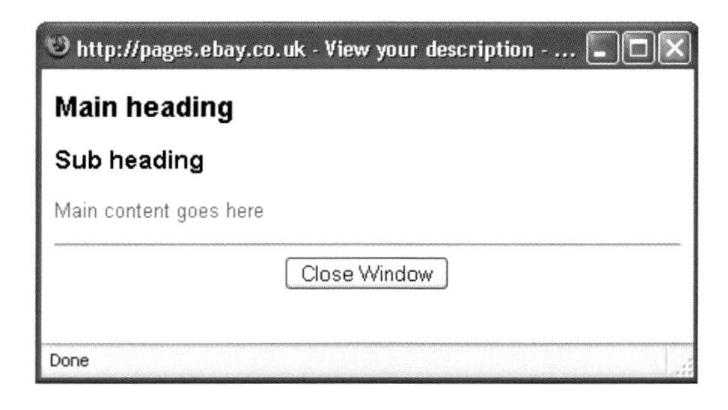

Font size

1. Add a <font size> </font> opening and closing tag to create a text size

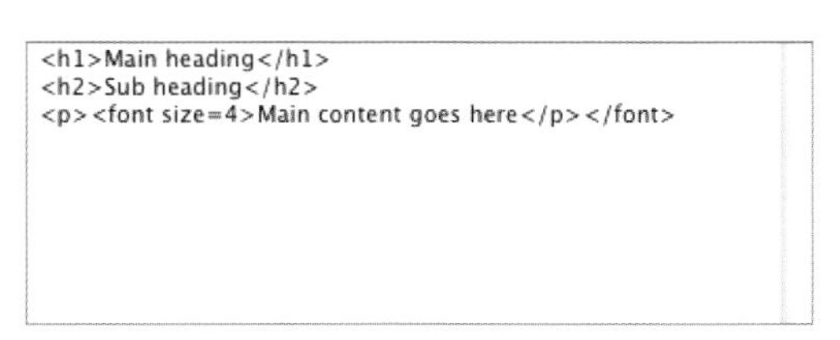

```
<h1>Main heading</h1>
<h2>Sub heading</h2>
<p><font size=4>Main content goes here</p></font>
```

2. Click Preview Description to see how the effect will look

Hot tip

The <font> tag can have different attributes applied to it, such as face (for a different font), size and colour. For each of these the attribute has to have a property assigned to it, for example <font size=4>.

...cont'd

Size and colour

1. Add a combination of font tags to change the size and colour of text

```
<h1>Main heading</h1>
<h2>Sub heading</h2>
<p><font size=4 color=red>Main content goes here</p></
font>
```

2. Click Preview Description to see how the effect will look

Hot tip

The <font size> tag and attribute can be used to create text that is larger than the <h1> tag, which appears at a preset size.

Centre

1. Add a <center> </center> opening and closing tag to centre text

```
<h1>Main heading</h1>
<h2>Sub heading</h2>
<p><font size=4 color=red><center>Main content goes
here</p></font></center>
```

2. Click Preview Description to see how the effect will look

Bulleted list

1. Add <ul><li></li></ul> opening and closing tags to create a bulleted list

```
<h1>Main heading</h1>
<h2>Sub heading</h2>
<p><font size=4 color=red><center>Main content goes
here</p></font></center>
<ul>
<li>List item 1</li>
<li>List item 2</li>
<li>List item 3</li>
</ul>
```

2. Click Preview Description to see how the effect will look

Don't forget

The code for a bulleted list consists of the <ul> tag (which stands for Unordered List), and the <li> tag (which stands for List Item) for each bullet point.

Numbered list

1. Add <ol><li></li></ol> opening and closing tags to create a numbered list

```
<h1>Main heading</h1>
<h2>Sub heading</h2>
<p><font size=4 color=red><center>Main content goes
here</p></font></center>
<ol>
<li>List item 1</li>
<li>List item 2</li>
<li>List item 3</li>
</ol>
```

2. Click Preview Description to see how the effect will look

Don't forget

The code for a numbered list consists of the <ol> tag (which stands for Ordered List), and the <li> tag for each numbered item.

Horizontal rule

1. Add an <hr> tag to create a horizontal rule

```
<h1>Main heading</h1>
<h2>Sub heading</h2>
<p><font size=4 color=red><center>Main content goes
here</p></font></center>
<hr>
<ol>
<li>List item 1</li>
<li>List item 2</li>
<li>List item 3</li>
</ol>
```

2. Click Preview Description to see how the effect will look

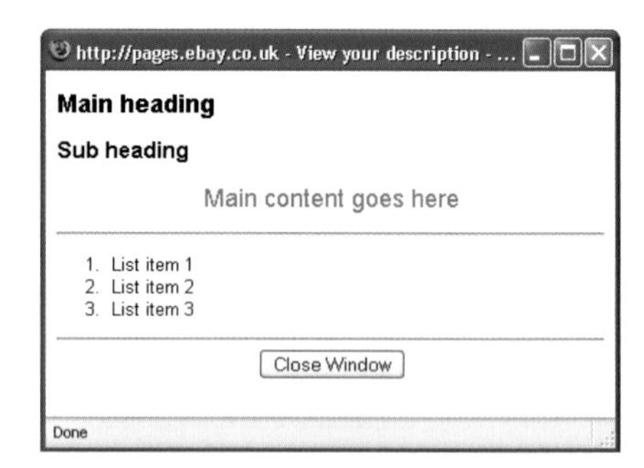

Don't forget

There is no closing tag for the <hr> tag.

Don't forget

The code for a table consists of <table> for the start of the table, <tr> for a new row, and <td> for each cell in a row. Each tag has to have a subsequent closing tag. Additional formatting can also be achieved with the Border and Width attributes.

Tables

1. Add <table> <tr><td></td> </tr></table> tags to create a table

```
<h1>Main heading</h1>
<h2>Sub heading</h2>
<p><font size=4 color=red><center>Main content goes
here</p></font></center>
<table border=1 width=100%>
<tr><td>One</td><td>Two</td></tr>
<tr><td>Three</td><td>Four</td></tr>
</table>
```

2. Click Preview Description to see how the effect will look

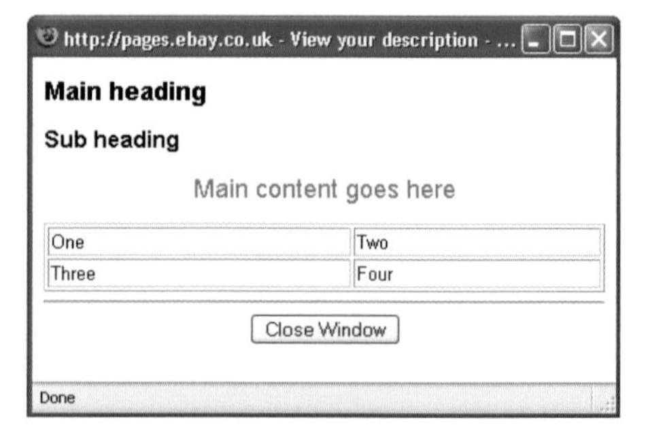

About Me page

The About Me page is an opportunity to go in for a bit of self-promotion within eBay. It is your own personal Web page that you can use to tell other eBayers a bit about yourself, your hobbies and the types of items that you like buying and selling. All registered eBay users can create their own About Me page and it is free to do so. Once you have created an About Me page a new icon appears next to your user ID. To create an About Me page:

1. Access the About Me authoring page either from the People section of the Community site or via the About Me link in the eBay sitemap, which can be accessed from the main navigation bar

Connect

- Find a Member
- UK & IE Discussion Boards
- eBay Groups
- About Me

2. The About Me introduction page explains a bit more about what the page does and the advantages of having one. Click the Create Your Page button

About Me
Tell the world about you and your interests
Talk about who you are
You'll soon find other people with similar interests in eBay's community.
Describe your collecting or selling passions
Relate how you began collecting for fun or selling for profit.
Build trust and confidence among potential trading partners
Highlight your trading record to reassure potential buyers or sellers.
Get your own personilised eBay page with the me icon
Send others to http://members.ebay.co.uk/aboutme/nickvan1/
Create your About Me page
Begin by clicking the button below.
Create Your Page

3. Click here to use the About Me wizard to create the page. Click Continue

The information on the About Me page should be as relevant as possible to your eBay activities.

...cont'd

4. Enter a page title and any information about yourself that you want to appear on the About Me page

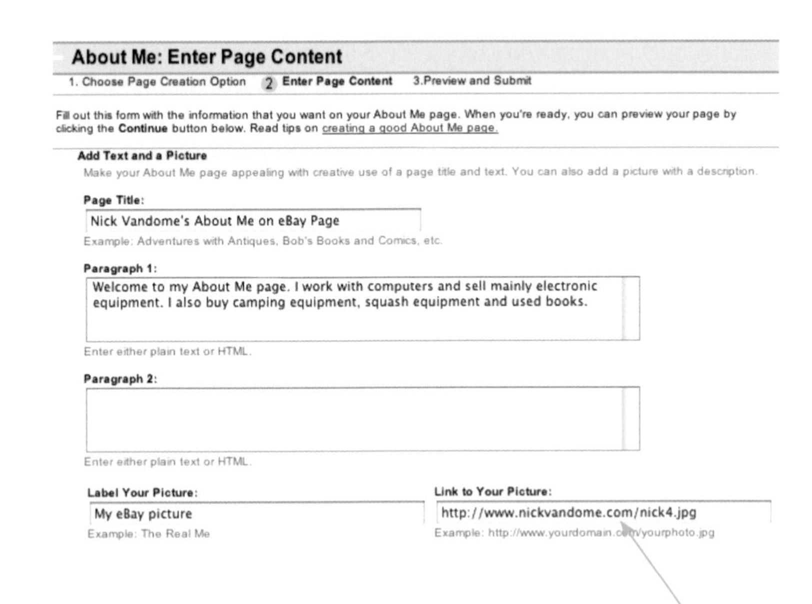

About Me: Enter Page Content

1. Choose Page Creation Option 2 **Enter Page Content** 3.Preview and Submit

Fill out this form with the information that you want on your About Me page. When you're ready, you can preview your page by clicking the **Continue** button below. Read tips on creating a good About Me page.

Add Text and a Picture

Make your About Me page appealing with creative use of a page title and text. You can also add a picture with a description.

Page Title:

Nick Vandome's About Me on eBay Page

Example: Adventures with Antiques, Bob's Books and Comics, etc.

Paragraph 1:

Welcome to my About Me page. I work with computers and sell mainly electronic equipment. I also buy camping equipment, squash equipment and used books.

Enter either plain text or HTML.

Paragraph 2:

Enter either plain text or HTML.

Label Your Picture:

My eBay picture

Example: The Real Me

Link to Your Picture:

http://www.nickvandome.com/nick4.jpg

Example: http://www.yourdomain.com/yourphoto.jpg

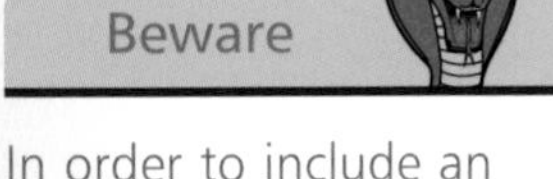

In order to include an image on your About Me page you have to have one already hosted on another website.

5. Enter a URL for any images you want to include on the page and give the picture a title

6. Select whether you want your feedback displayed on your page and select any website links you want included

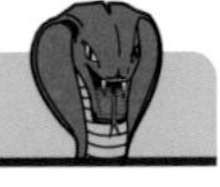

If you have any negative feedback it is not a good idea to have this displayed on your About Me page.

Show Your eBay Activity

You can include your latest listings and feedback information. If you have more than 100 listings or a feedback score greater than 200, eBay will automatically include a link to your Member Profile page.

Show Feedback You've Received:

Show 10 most recent comments

Label Your Listings:

Example: P.G. Wodehouse First Editions

Show Your Current Listings:

Show no items

Add Links

You can point to your favorite Web addresses and give names to the links. Read the About Me Page policy.

Name:

Nick Vandome's website

Example: All About Movies

Web Address:

http://www.nickvandome.com

Example: http://www.imdb.com

Name:

Web Address:

http://

Name:

Web Address:

http://

< Back   Continue >

7. Select a format for your page and click Submit to preview it

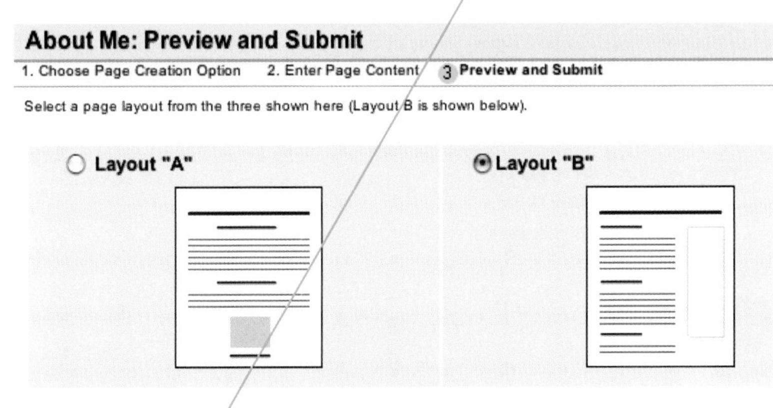

The Preview page may look slightly different from the final published page, particularly some of the font sizes.

8. Preview the page and, if you are happy with it, click Submit again to have your About Me page uploaded onto the eBay site

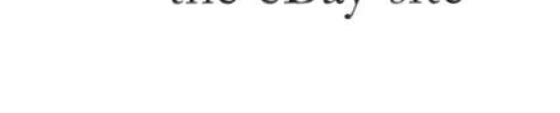

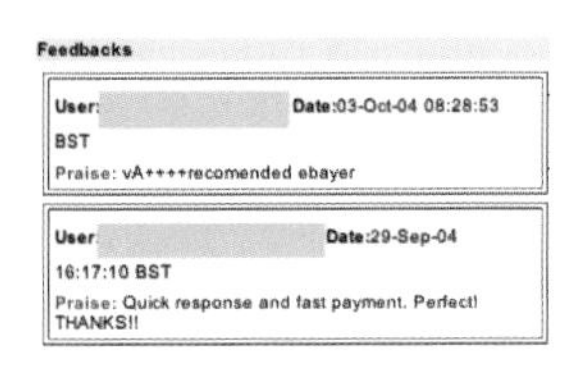

...cont'd

9. A confirmation message appears stating that your About Me page has been created. Click here to view the page on eBay as other users will see it

Your New About Me Page Has Been Created

Your new About Me page has been successfully created. You should shortly be able to view it at this Web address: http://members.ebay.co.uk/aboutme/nickvan1.

Where would you like to go next?

- Item I last looked at
- My eBay
- More Community Resources

10. Preview the live About Me page

About Me: nickvan1 (7) me

Nick Vandome's About Me on eBay Page

My eBay picture

Welcome to my About Me page. I work with computers and sell mainly electronic equipment. I also buy camping equipment, squash equipment and used books.

Favourite Links

Nick Vandome's website

Feedbacks

User:baza100 (458) Date:03-Oct-04 08:28:53 BST

Praise: vA++++recomended ebayer

User:chinaebase (3900) Date:29-Sep-04 16:17:10

11. Once you have created an About Me page, a new icon appears next to your User ID. This applies to all users who have created their own About Me page

Seller information

nickvan1 (7) me

Feedback Score: 7

Positive Feedback: 100%

Member since 11-Aug-04 in United Kingdom

Read feedback comments

Ask seller a question

View seller's other items

Standard Purchase Protection Offered.
Find out more

Hot tip

Have a look at the About Me pages of other eBay users to see what they have to say about themselves, and the style of their pages.

Emphasising items

When listing items for sale you can boost your chances of a successful sale by emphasising your item with either bold or a highlighted background. There is a small additional listing fee for this, but it can be a worthwhile investment.

Adding bold

To add bold to a listing:

1. On the listing page, tick this box

Add Gallery above in pictures
Bold (£0.75)
Attract the attention of buyers - use **Bold**.

2. Once the item is listed, the text appears in bold to make it stand out more next to other listings

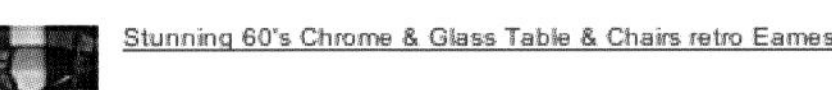

> **Hot tip**
>
> Some sellers use all capitals for the titles of items. This can make them stand out more in search results pages.

Adding highlights

To add a highlighted background to a listing:

1. On the listing page, tick this box

Highlight (£2.50)
Make your listing stand out with a coloured band

2. The whole listing appears on a coloured background

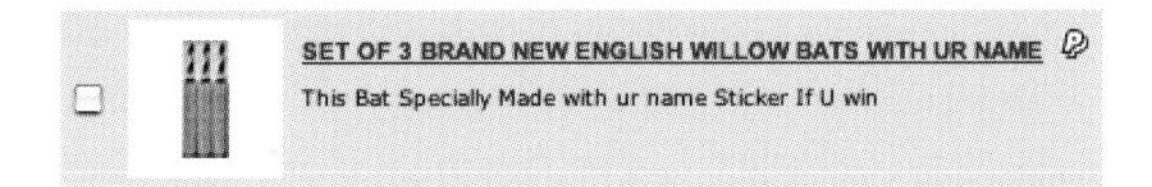

Featured items

Another way of improving your sale chances is to choose to have your item featured in certain parts of eBay. This means that your item will appear in more prominent places within the site: either at the top of search results pages or on the eBay home page itself.

There are additional listing fees to have items featured in this way, and in most cases it is probably only worthwhile for items which are going to fetch reasonably high prices. The current fee to have items featured at the top of search results pages is £9.95 (Featured Plus!) and to have them on the home page is £49.95. Therefore it is not worth doing this for items which you think will only fetch a few pounds in an auction. However, for higher priced items it can be a good way to attract more potential buyers and therefore increase the chance of an active and successful auction. To list featured items you have to have an overall feedback rating of 10 or over. Once you have reached this level, the featured items option will be available within the listings page. Until then it will not be available, but there will be a link giving you more information:

If you are using eBay to conduct an online business it may be worth assessing whether the additional fee of having a featured item would be cost-effective in terms of overall profit.

Viewing Featured Plus! items

To view items that are featured at the top of search results:

1. Click on a category item, or enter a keyword into the search box

2 The featured items are displayed at the top of the search results page. By default, they appear with the auctions ending soonest at the top

1485 items found in **Elizabeth II (1952-Now)** ▪ Add to My Favourite Categories ▪ Sell in this category

List View | Picture Gallery **Sort by:** Time: ending soonest Customise Display

Compare	Item Title	Bids	Price	Postage	PayPal	Time Left ▲
Featured Items						
☐	Silver Jubilee Commemorative plates. Queen Elizabeth 2 Queen Elizabeth II, 1977. Royal Grafton	-	**£2.99**	£2.50		10h 01m
☐	**Carlsberg Royal Lager To Celebrate Coronation-Uno pened!**	-	**£120.00**	£20.00		5d 06h 10m
☐	QUEEN ELIZABETH II 80th BIRTHDAY LTD. EDTN "LOVING CUP" 80th BIRTHDAY OF HER MAJESTY - 21st APRIL 2006	Buy It Now	£12.50	£3.00		7d 01h 07m

3 Scroll down the page to see the non-featured items. There is normally a line that divides the featured and non-featured items

Optimise your selling success! Find out how to promote your items

	Item Title	Bids	Price	Postage	PayPal	Time Left
☐	a pair of silver jubilee cup and saucer	-	**£0.99**	£5.95		1m
☐	Coronation Queen Elizabeth 2 commemerative glass beaker	-	**£0.99**	£3.00		26m
☐	4 coasters in a tin from buckingham palace	-	**£3.00**	£1.30		46m
☐	WEDGWOOD PLATE - GOLDEN JUBILEE	-	**£0.99**	£3.00		49m
☐	SET 5 PHOTO CARDS EARLY ROYAL FAMILY	-	**£0.99**	£0.90		49m

Non-featured items can still have bold and highlighting added to them for emphasis.

...cont'd

Viewing home page featured items

Items that are featured on the eBay home page are displayed in a separate box:

1. Click on an item to go directly to that auction

2. Click here to view all of the home page featured items

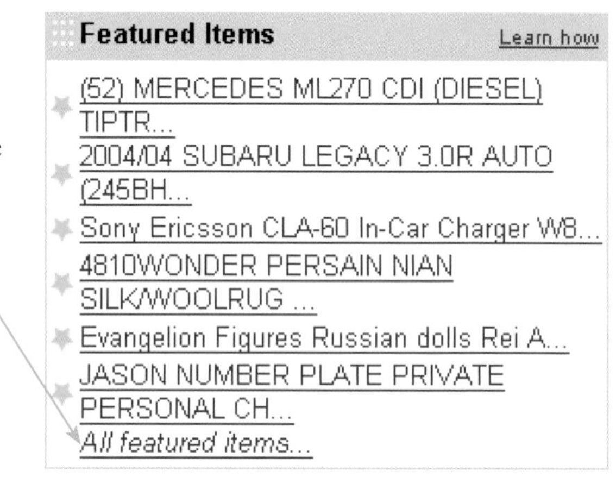

Home page featured items rotate in the home page box, because there are too many of them to be displayed at once.

3. All of the home page featured items are listed. By default, they appear with the auctions ending soonest at the top

Featured Items

Item	Bids	Price		Time left
CAESAR BED KING SIZE MAHOGANY * NEW *	14	**£195.02**	£80.00	2h 01m
THE ULTIMATE KART KARTING domain name	-	**£500.00**	--	3h 08m
Restored Lambretta 200 cc GP's 6 Month Warranty In Stock See B4 U Buy !!	- Buy It Now	**£1,900.00** £1,995.00	£100.00	3h 10m
MERCEDES CL500 (S Class) COUPE AUTOMATIC 1995 Rare, superb condition with LOW RESERVE	15	**£4,000.00**	--	3h 36m

8 Conducting an auction

Once an auction starts it is important to monitor and manage it, to deal with any problems that may occur and to maximise the amount of money from the sale.

Once an auction starts

When an auction in which you are selling something starts, you can just sit back and wait until the end of the auction to see what happens and what price has been achieved for the item. However, for a more successful eBay career it is important to keep an eye on the auction to see how it is progressing and, if necessary, give it some added impetus. Initially there are two things that you should do once you have completed listing an item and the auction has started:

1. Conduct a search for the item to make sure it has been registered in the eBay database of items for sale. This can take a few hours, since the database is updated throughout the day. The item should have a sunrise icon next to it, indicating that it is a new listing. Click on the link for your item to view the actual auction page

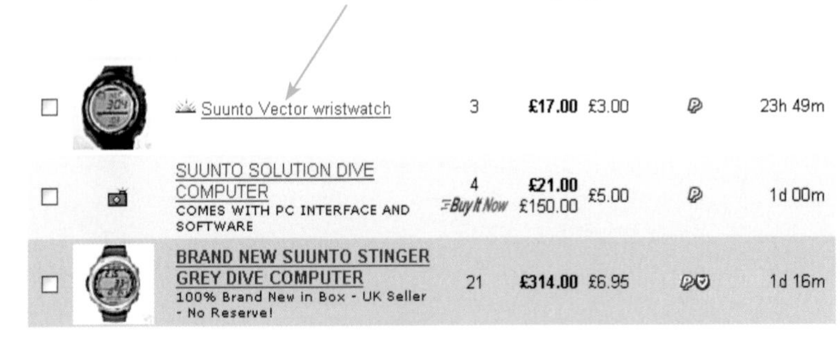

2. Check on the auction page to make sure the listing and image are as they should be. You can also check here periodically to see the number of bids that have been made (this will also be visible on your My eBay page)

Current bid: £17.00
Place Bid >
Time left: 23 hours 46 mins
1-day listing, Ends 27-Mar-06 21:24:01 BST
Start time: 26-Mar-06 21:24:01 BST
History: 3 bids (£5.00 starting bid)
High bidder: edabcd (141 ★)
Item location: Perth
United Kingdom
Post to: United Kingdom, Ireland
Postage costs: £3.00 -- Royal Mail 1st Class Standard
Other postage services available
Larger Picture
Postage, payment details and return policy

Don't forget

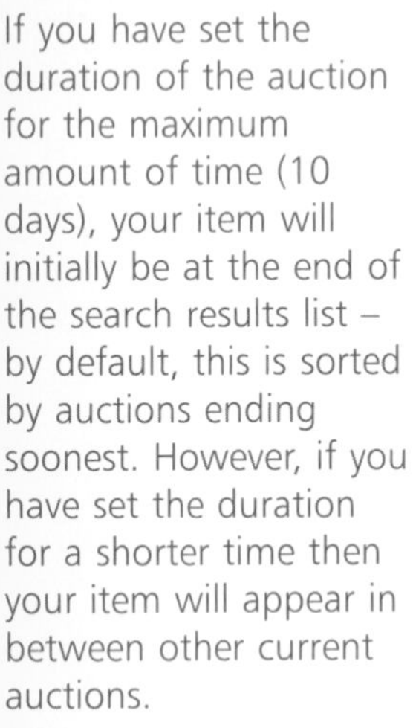

If you have set the duration of the auction for the maximum amount of time (10 days), your item will initially be at the end of the search results list – by default, this is sorted by auctions ending soonest. However, if you have set the duration for a shorter time then your item will appear in between other current auctions.

Answering questions

It is important to be available during an auction so that you can answer any questions that prospective buyers might have. These could be anything from asking about the postage arrangements, to requesting more information about the technical specification of an item. A good seller will always try to answer questions as quickly and as honestly as possible. If you leave a question unanswered then the prospective buyer will probably not proceed with a bid.

1. When a prospective buyer asks a question, it will be flagged up on the auction page (which only you can see) and you will also be sent an email with the question

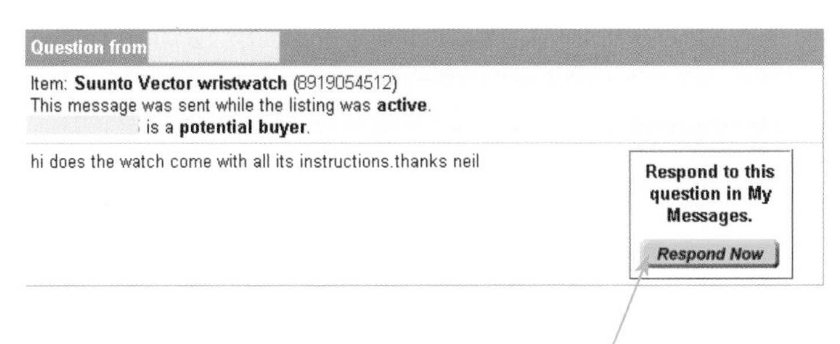

2. Click the Respond Now button to reply to the question

3. Enter your reply in the box underneath the question

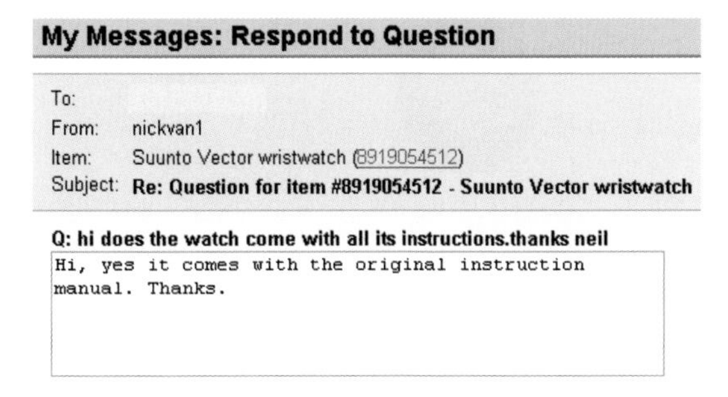

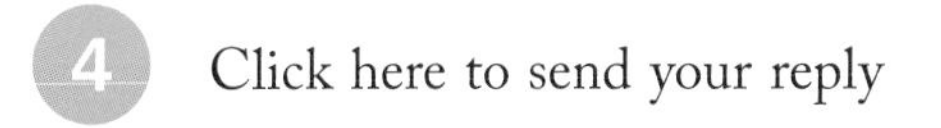

4. Click here to send your reply

If you have several auctions on the go at the same time make sure you answer questions in relation to the correct one.

Promoting items

If an auction for a particular item is not going well – for example, if you have had no bids for it – you can still return to the original listing page and add some promotional functions, to try to boost the item's appeal. To add promotional items to an existing auction:

Open the relevant auction page and click the 'Promote your item' link

Suunto Vector wristwatch

You are signed in

Seller status: Your item has been bid up to £17.00

Change your cross-promoted items

Revise your item

Promote your item

Sell a similar item

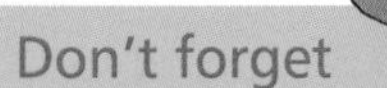

The options for promoting an existing auction are the same as those for a listing when it is first made.

2 Select any options for promoting your auction and click Continue at the bottom of the page

Gallery Picture
Please choose from the pictures below, and select a gallery option.

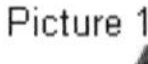

✓ Gallery (£0.15)
Add a small version of your first picture to Search and Listings. See example

☑ Bold (£0.75)
Attract the attention of buyers - use **Bold**. See example

☐ Highlight (£2.50)
Make your listing stand out with a coloured band in Search results. See example

Revising listings

As well as adding new promotional details to an existing auction, it is also possible to edit some of the details, such as the title, the description, and the postal charges, or to list the item under a second category. Revisions can be made if there have been no bids for an item and there is more than 12 hours left until the end of the auction. If your item has received a bid, or there is less than 12 hours left, then you can only add to the description or add promotional items. To revise a listing:

Adding a listing to an additional category means that there is more chance of users finding it during a search. However, you will have to pay two insertion fees, one for each category.

1. Open the relevant auction page and click the 'Revise your item link'

Suunto Vector wristwatch

You are signed in

Seller status: Your item has been bid up to £17.00

You have 1 question to answer
Change your cross-promoted items
Revise your item
Promote your item
Sell a similar item

2. Click here to see how your listing will look

Step 1: Review your listing

Listed on: 26-Mar-06 21:20:37 BST
Suunto Vector wristwatch (#8919054512)
Current status: Your listing has bids/sales or has less than 12 hours left
Preview how your item will look to buyers

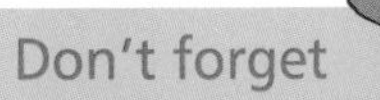

Once an item has been revised, this is noted on the auction page, to alert potential buyers to this fact.

3. Scroll through the listing until you find the element to change

4. Click here to edit a particular element of a listing

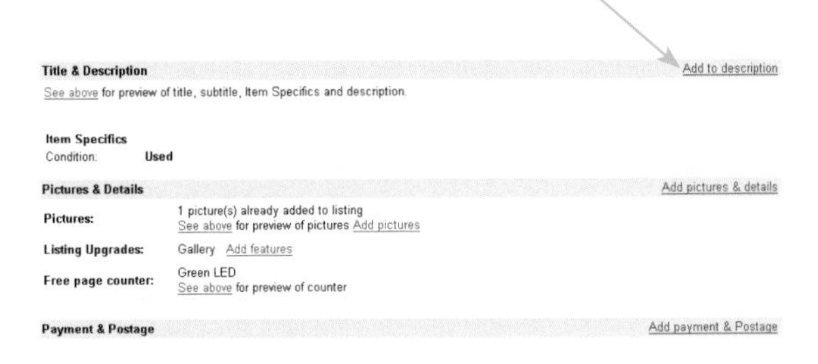

...cont'd

5. Click one of the Edit button to edit a particular element of a listing

Edit title

Edit subtitle

Edit description

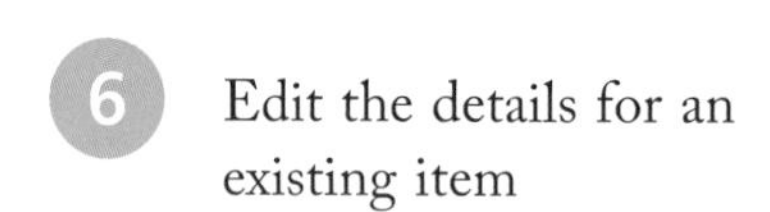

6. Edit the details for an existing item

Postage *

Specify a flat cost for each postal service you offer.

Domestic Postage (offer up to 3 services)

	£
Royal Mail 1st Class Standard (1 to 2 working days)	5.00
Royal Mail 1st Class Standard (1 to 2 working days)	4.00
Select a postage service	

7. Click Save Changes

Cancel Changes | Save Changes

8. Click Submit Revisions to update the existing listing

Cancel Revision | Submit Revisions

Don't forget

If a prospective buyer has asked a question about a listing, such as the amount for postage, this could be used as the basis for reviewing and revising the listing. However, make sure the request is reasonable.

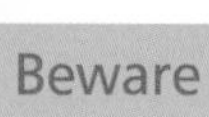

Beware

If you agree to change a listing in response to a request from a user, make sure you change the actual listing rather than agree the change directly with the person who has made the request.

When an auction finishes

When an auction of your own is in progress it can be a nerve-wracking process just waiting to see if you get any bids at all for your item. As long as you get at least one bid (and you have not put a reserve price on the item) you will be able to sell the item to the highest bidder at the end of the auction. Once this happens you will have to perform a few tasks to help the buyer pay you so that they can receive the item:

1. Once the auction finishes, you will be sent an email notifying you of the winning bidder. Click here to send them payment details

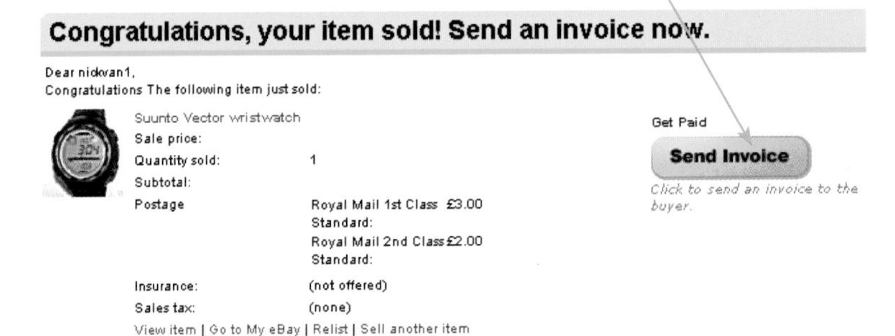

2. Alternatively, go to the auction page and click here

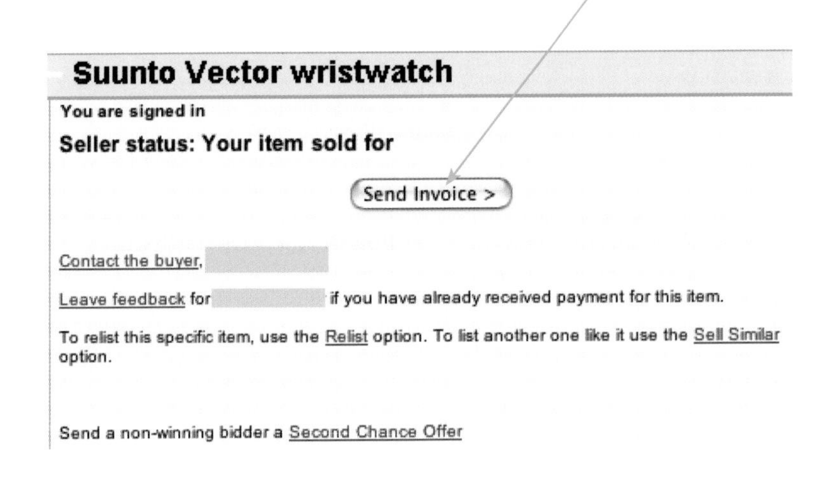

Don't forget

If an item has no bids, and therefore no buyers, you will have the chance to relist it for sale. This will involve paying the listing fees again. If you do relist an item, take a critical look at the original listing to see if there are any ways in which it could be improved.

Don't forget

If the person who has won an auction does not pay you or backs out of the transaction, you can offer it to the second-highest bidder. This is known as a Second Chance Offer.

...cont'd

As well as an invoice sent by the seller, a buyer will also be sent payment details by eBay.

3. The invoice is generated and sent to the buyer

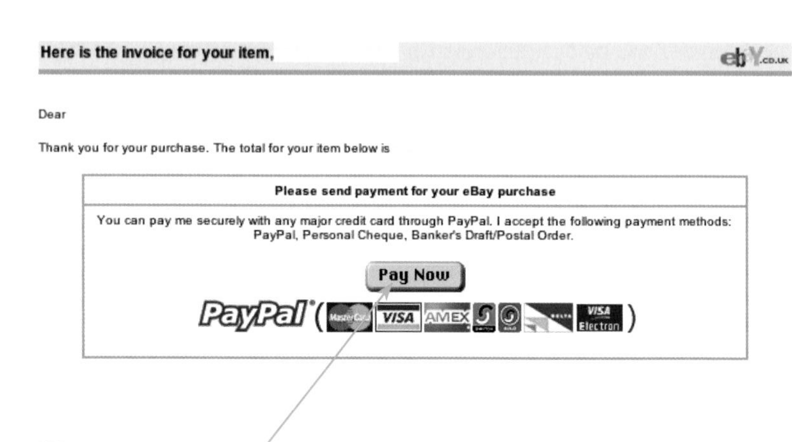
Here is the invoice for your item,

Dear

Thank you for your purchase. The total for your item below is

Please send payment for your eBay purchase

You can pay me securely with any major credit card through PayPal. I accept the following payment methods: PayPal, Personal Cheque, Banker's Draft/Postal Order.

Pay Now

4. The buyer can pay by clicking the Pay Now button

5. When the item has been paid for, you will be sent confirmation of this and it will also be displayed when you access the auction page for the item

Suunto Vector wristwatch

This item has been paid for through PayPal. Payment was sent to: on 27-Mar-06.

Mark as Dispatched >

If you have shipped this item, click the **Mark as Dispatched** button.

Other actions for this item:

You can manage all your items in My eBay and do the following:

- Leave feedback for this item.
- View payment details.
- View PayPal payment for this item.
- Contact the buyer, about this item.

Receiving payment

From the seller's point-of-view, the most important part of a transaction is getting paid. In some cases the buyer will just send the payment once they have been notified that they have won the auction. If this happens, it is just a matter of waiting until you receive the payment before sending off the item. In other instances, the buyer may contact you to ask exactly how you would like payment:

1. If a buyer sends an email enquiring about payment, you will be able to give them exact details by replying to the message

From:
Subject: **ref: eBay purchase - item #3845558689, Nikkor AF 35-70mm lens**
Date: October 17, 2004 12:41:54 AM GMT+01:00
To: Nick Vandome

Hello

How would you like me to pay? Cheque or PayPal?

Regards

2. If you are receiving payment via PayPal, you will be sent an email confirmation once the funds have been paid

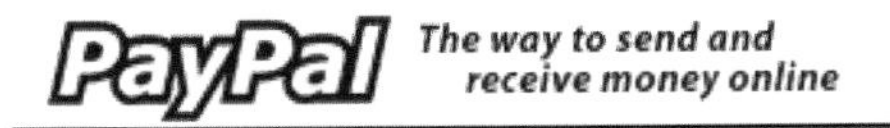

PayPal
The way to send and receive money online

Dear Nick Vandome,

This email confirms that you have received a 46.50 GBP from

View the details of this transaction here: 51Y817318L166843H

Don't forget

If you have a PayPal account and you receive a credit card payment to it, you will need to upgrade to the Premier PayPal account in order to accept and access this payment. This is looked at in more detail in Chapter Ten.

Posting items

The two golden rules about posting items are:

- Take the items to a post office and get them weighed before you list them for sale. This way you will not get a nasty shock at the time of posting. If you have to pay more for the postage than you charged the buyer then it is simply a waste of money. Conversely, do not overcharge for postage: find out the correct amount and then charge the buyer this, or slightly more to take into account the cost of packaging and your own time.
- Package items in a way in which you would like to receive something yourself. If the buyer sees that you have made an effort with the packaging then they will be more likely to leave you positive feedback and also to buy from you again.

Depending on the type of item you are posting, there are different alternatives for packaging. If in doubt, use more packaging rather than less.

- Brown paper and string. If you are sending non-breakable items, such as clothes, then one option would be to package them as a parcel with brown paper and string. If you do this, make sure there are at least a couple of layers of paper, in case one layer gets torn.
- Padded envelopes. There are a variety of padded envelopes available at most post offices. These are ideal for items such as books, CDs and DVDs.
- Boxes. Different sized boxes can be bought at the post office, but it is also a good idea to collect boxes, such as shoeboxes, for use when sending eBay items to buyers.
- Bubble wrap. This can be bought in a large roll and it is ideal for wrapping items and also packing boxes.
- Additional packaging. Other useful items for packing boxes include newspaper and the polystyrene that comes in packaging for items such as televisions and computers.

Don't forget

Although books, CDs and DVDs can be sent in ordinary envelopes, it shows the buyer that you have made the extra effort if you send them in a padded envelope.

Hot tip

The Royal Mail has a variety of comprehensive delivery solutions for both business and personal customers. These include the delivery of bulky or expensive items. For full details, take a look at their website: **www.royalmail.com**

Hot tip

Courier services are also a good option for delivery of items. Some to look at are: DHL at **www.dhl.co.uk** CitiPak at **www.citipak.co.uk**, Parcel2go at **www.parcel2go.com**

9 Feedback

Feedback from eBay users is vital to the smooth running of the site. This is used to assess how trustworthy, or otherwise, users are and it is an invaluable tool in the self-regulation process.

Why feedback matters

Unlike a lot of websites where there is a rarely used option for leaving feedback or comments, the feedback mechanism on eBay is vital to the continuing success of the site. Feedback is actively encouraged from both buyers and sellers for all transactions. The amount of feedback appears next to a User's ID and all other registered eBay users can access this feedback to see what has been said about someone.

It is possible to block your feedback from other users, but this is not recommended as it suggests you have something to hide. You can block your feedback by visiting the Help section within the Community pages. Then type Feedback Private into the search engine and you will be taken to a page where you can hide your feedback. However, this only applies to the feedback comments, not your overall feedback profile. Always think carefully before dealing with anyone with feedback that has been hidden.

We all like to hear what other people think about us but on eBay feedback is a lot more than idle curiosity: it informs buyers and sellers about people's track records on eBay and how they perform when they are conducting transactions. In effect, it is an extremely effective method of self-regulation by eBay users. If a buyer or a seller is acting in an unprofessional or dishonest way then this will quickly show up in the feedback, and their eBay career may come to a swift end. Some points to remember about feedback:

- There are three types of feedback that can be given: positive, neutral and negative.
- Feedback stays with you throughout your eBay career. There are certain circumstances when feedback can be removed, but in general it sticks to you like glue.
- Feedback is usually available for all other users to read and evaluate. Your feedback is visible to everyone else and you can check the feedback of everyone else too.
- You can give feedback only after you have completed a transaction (bought or sold something).
- Feedback should be honest, accurate and informative. Essentially, feedback is for the benefit of other eBay users, so they can decide if they want to deal with this user. Try to make your feedback worthwhile and informative.
- When making a transaction with another eBay user always, always read the feedback that has been left about them from other users.

Types of feedback

Positive feedback

This is the most common type of feedback and is given by both a buyer and a seller if a transaction goes smoothly: when the buyer pays promptly, the seller sends the goods promptly and they are just as they are described in the auction. This type of feedback makes up the vast majority that is left on eBay. Technically, a seller should be in a position to leave feedback before a buyer, as they will receive payment before the goods are despatched. In reality, some sellers wait until they have received feedback from the buyer before leaving some of their own.

Positive feedback can be left by both a buyer and a seller.

Neutral feedback

This is the least common type of feedback and is left when either a buyer or a seller is not entirely happy with a transaction. It might be because one element of the deal was unsatisfactory, such as a description of an item being slightly inaccurate, while the rest of it was conducted as expected. This is a bit of a grey area on eBay and, in reality, most users have fairly strong opinions when it comes to feedback: they either like the way a transaction has been handled or they don't. However, neutral feedback comments should be read carefully to see what the issue was and whether you feel this would effect your dealings with the person involved.

Negative feedback

This is the most serious type of feedback as it indicates a problem with a transaction. This could be for a variety of reasons; from payment not being sent, to fraud where a seller accepts payment for an item and then does not send it. In cases like this, the amount of negative feedback generated would ensure the person involved would not prosper on eBay for long. However, until the feedback appears they may still get away with it for a few transactions. Negative feedback is a serious issue on eBay and you should always think twice before you leave any, and strive to ensure that you have none left for yourself.

Hot tip 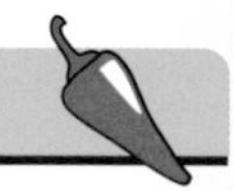

A good way to quickly build up positive feedback is to buy cheap items that can be delivered online. An ideal area for this is recipes. If the seller accepts PayPal, then the whole operation can be completed within a few minutes, including the receiving and leaving of feedback.

Don't forget 

For more on negative feedback, see pages 153–155.

Accessing feedback

It is possible to access feedback for all registered eBay users and also to view the feedback that has been left about you by other people. To view other users' feedback click on their User ID wherever it appears, such as on an auction page. You can view your own feedback through your My eBay page or via your User ID.

Viewing feedback on the My eBay page

1. Open your My eBay page and click the Feedback link to access your own feedback

My Account

- Personal Information
- Addresses
- Preferences
- Feedback

It is not obligatory to leave feedback, so don't be offended if someone does not leave you any even after you have left some for them.

2. This icon denotes that feedback has been left for a particular item

3. The most recent feedback is displayed on your My eBay page

Recent Feedback (View all feedback)

Comment	From		Date/Time	Item #
great ebayer	Seller:		13-Feb-06 11:54	7217675970
good transaction	Buyer:		01-Jul-05 06:30	4552528603
Great!. A pleasure to do business with. Thank you very much!	Seller:		28-May-05 22:56	7517255253
FAST PAYMENT !!! GOOD BUYER !!!	Seller:		27-May-05 17:41	5775826599
Great buyer...prompt payment, valued customer, highly recommended.	Seller:		25-May-05 11:17	4382942094

Viewing feedback via User ID

To view all of your feedback via your User ID:

1. Click here on your own User ID

Hello, nickvan1 (19 ☆) me

2. The current feedback score is shown here

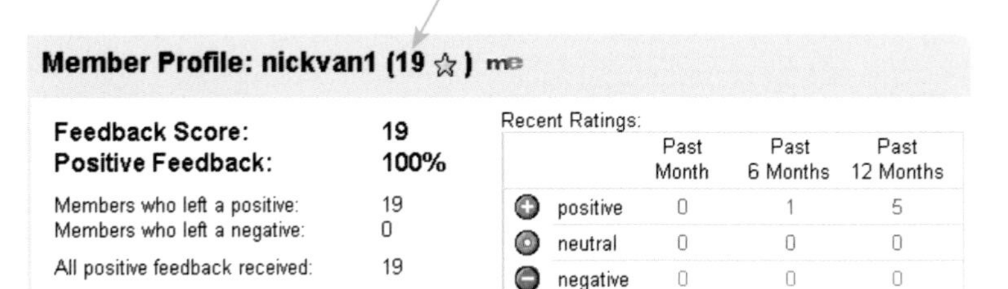

3. A summary of feedback for the last 12 months is shown here

4. Scroll down the page to see all of the feedback comments

Feedback Received | From Buyers | From Sellers | Left for Others

19 feedback received by nickvan1 (0 ratings mutually withdrawn)

Comment	From
great ebayer	Seller
good transaction	Buyer
Great!. A pleasure to do business with. Thank you very much!	Seller
FAST PAYMENT !!! GOOD BUYER !!!	Seller
Great buyer...prompt payment, valued customer, highly recommended.	Seller no long
Item delivered speedily and in good condition. Good contact too.	Buyer
good value ,good fast delevery ,hope to buisness again.thanks	Buyer
great buyer, fast payment, good transaction - thanks	Seller

Hot tip

Viewing your own feedback via the User ID is just the same as viewing the feedback for another user.

Don't forget

The feedback score is the aggregate of positive and negative feedback. The positive feedback figure is a percentage of positive feedback in relation to all of the feedback you have received.

Don't forget

Click on the tabs along the top of the feedback window to view feedback from just buyers, just sellers or feedback you have left for others.

Giving feedback

Whenever you are involved in a transaction it is good manners to leave feedback for the other person. This is not only polite, it also helps the rest of the eBay community assess the performance of the other person. To leave feedback:

Don't forget

Feedback reminders are automatically placed on your My eBay page once you have completed a transaction.

1. Go to your My eBay page and click on the Leave Feedback button

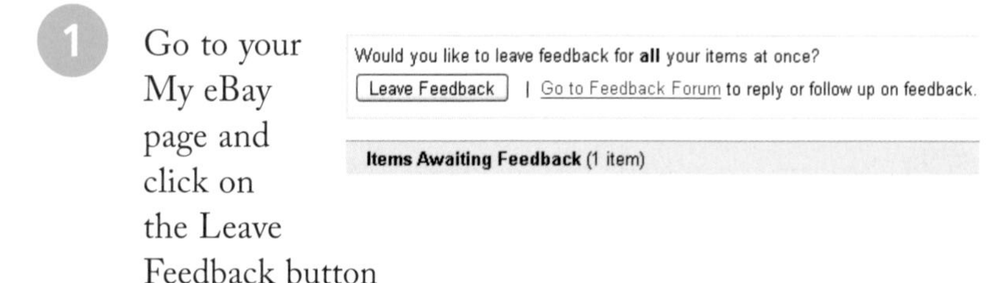

2. Alternatively, go to the Items I've Won section on your My eBay page and click here

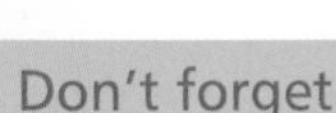

Don't forget

You can also leave feedback by going to the original auction page once it has finished and following the feedback link there.

3. Use these buttons to select the type of feedback to leave

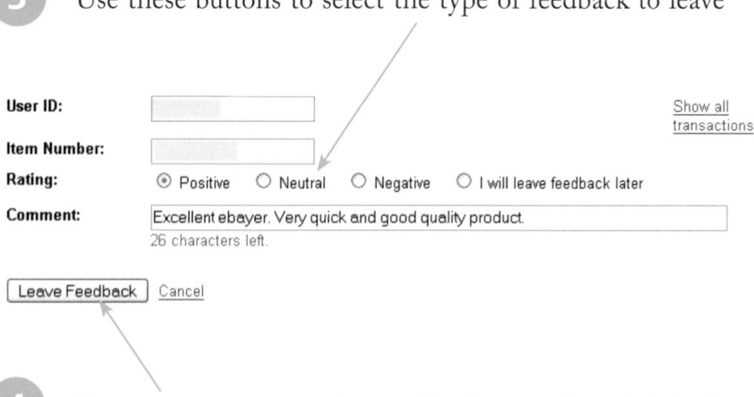

4. Enter your comments (up to 80 characters) and click the Leave Feedback button

Negative feedback

Since most of eBay's business is based on trust between the users, negative feedback is a serious issue. It can result in prospective buyers steering clear of a particular auction and, ultimately, ruining someone's eBay career. Negative feedback is like a very public black mark against your name. In many cases this can be justified – reasons for leaving negative feedback include:

- Buyers not paying once they have won an auction.
- Sellers not sending items once they have received payment.
- Goods being delivered but being nothing like their description in the auction.
- Rude or obstructive behaviour when a buyer has tried to get more information from a seller.

However, there are some cases where negative feedback is not always justified:

- Genuine mistakes. These do occur and negative feedback can be left as a result of an unavoidable mix up, such as an item being lost in the post.
- Mischief makers. Some people will deliberately leave negative feedback to inconvenience people. Since all users are allowed to leave feedback after a transaction has been completed there is not much that can be done about this, except to try to answer the feedback in a rational and even-handed way.
- Retaliatory negative feedback. Sometimes when a user leaves negative feedback the other person will do the same just out of spite (as long as they have not already left feedback). This is frequently unjustified but, again, there is not much that can be done about stopping it.

Try to reduce the chances of negative feedback by paying promptly, sending sold items quickly (in the described condition) and communicating effectively with users.

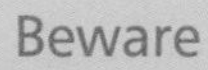

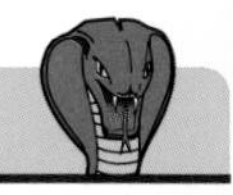

Never threaten another user with negative feedback as a means to gain an advantage in some way. This could get you banned from eBay.

Leaving negative feedback

Just as you should strive to not have any negative feedback left about yourself, so leaving negative feedback for others should be a last resort. If you are unhappy about a transaction, take time to evaluate the situation before giving negative feedback:

- If an item you have received is not how you think it should be, go back to the auction page and check the description, to make sure you have not missed anything.
- Check to see whether the person with whom you are dealing is a new eBay user. Some new users do not understand fully how the site works and this can result in misunderstandings.
- If an item that you have bought and paid for is not delivered, or no payment is sent for an item that you have sold, contact the other user. This can be done by clicking on their User ID to view their user profile and then clicking on the Contact Member button. Make sure you convey the exact nature of your concern.
- Leave a reasonable amount of time once you have contacted the other person.

Don't forget

If a user does not reply to your email queries, you could try the Square Trade mediation service, which is looked at on page 180.

If you have explored these avenues and you are still dissatisfied then you may have no choice but to leave negative feedback. Some people do not like doing this, as they feel it could result in retaliatory feedback from the other person; but, if you have a justified case, it is important that you do leave negative feedback. It is not a case of seeking revenge, it is about helping other eBay users come to a decision about whether to deal with this person or not. In some cases, you may be helping people avoid someone who should not be conducting business on eBay.

Negative feedback is left in the same way as leaving positive or neutral feedback. However, you are responsible for your comments, so make sure you are not abusive, defamatory or libellous. Make your point honestly and dispassionately, in a way that voices your concerns and alerts other users.

Replying to negative feedback

If you do receive negative feedback, for whatever reason, you are entitled to reply to it and it is important that you do so. Otherwise the negative comments will remain there unanswered and other users will think that you agree with the criticism and accept it. Once negative feedback is received, you will be alerted and given the chance to respond. However, before you do so, there are some points to bear in mind:

- Have a look at the user profile of the person giving the negative feedback. This will display the feedback that they have received and also the feedback they have left for others. You may discover that they are a serial leaver of negative feedback or that their own feedback rating is very poor. If this is the case you can mention it when you reply.
- Check the negative feedback to see if there is any truth in it. If there is, then be honest about this when you reply.
- If the negative feedback is retaliatory – in reply to negative feedback that you have posted about the other user – note this in your reply, as tactfully as possible.
- Leave it a day or two before you reply to negative feedback. The immediate reaction to being criticised is to fire off a strongly worded retort. However, it is better to look at it calmly.
- Since you only have one opportunity to reply, you should choose your words carefully so that they address the criticism rather than insult the other user.

The receipt of negative feedback can be unpleasant but it does not have to be disastrous. If you play fair and approach any criticism in the appropriate manner then some negative feedback should not have a detrimental effect on your eBay career. If you reply fairly and accurately to any negative feedback then you should not have too much to worry about in the long term.

Don't forget

You can reply to negative feedback only once, so make sure your response is as effective as possible, and try to avoid being too confrontational.

Beware

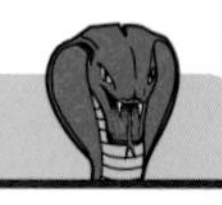

Never use abusive language when replying to negative feedback, regardless of what the other person has said about you.

Withdrawing feedback

In exceptional circumstances eBay will let you withdraw feedback, but only with the consent of the other user. You might withdraw feedback because of a genuine misunderstanding or if you left the feedback for the wrong person by mistake (however, if this happens, you have to leave the feedback for the correct person as soon as it has been withdrawn). To withdraw feedback:

1. Contact the other person involved and agree that you want to mutually withdraw an item of feedback

2. Go to the Help section within the Community pages and enter Feedback Withdrawal into the search engine. From the results, select the relevant page and click the Online Form link

Mutual Feedback Withdrawal

Mutual feedback withdrawal allows members to withdraw feedback for a transaction if both members are able to agree on a resolution. Withdrawn feedback is not counted in the feedback score or rating totals, but remains in the list of comments.

Once both members agree by completing the online form, feedback left by both parties is withdrawn at the same time. If a member has not left feedback for the transaction, by agreeing to mutual feedback withdrawal, they will give up the ability to leave feedback for the transaction.

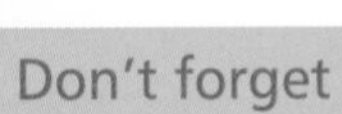

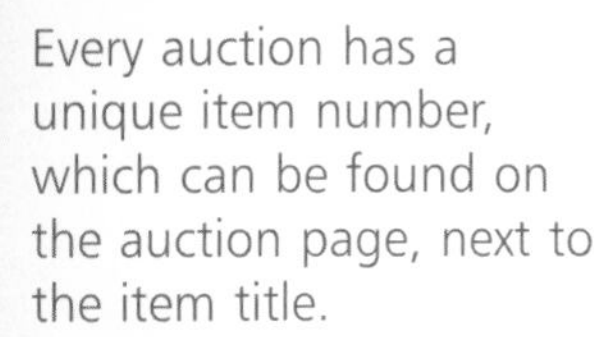

Every auction has a unique item number, which can be found on the auction page, next to the item title.

3. Enter the item number from which the feedback is to be withdrawn

Enter an item number:

1234

Continue >

4. Click Continue

5. Enter any comment and click Continue to allow the other user to complete their instructions

10 Financial matters

Sooner or later all eBay users will have to confront the financial side of operating on the site, whether it is as a buyer or a seller.

eBay fees

eBay makes its money by charging users to list items for sale and also by taking a percentage fee of the final sale price. The insertion fee is based on the starting price for an item, and the seller is responsible for paying both the listing fee and the final value fee. The current insertion fees are:

- £0.01 to £0.99, insertion fee is £0.15
- £1.00 to £4.99, insertion fee is £0.20
- £5.00 to £14.99, insertion fee is £0.35
- £15.00 to £29.99, insertion fee is £0.75
- £30.00 to £99.99, insertion fee is £1.50
- £100.00 and over, insertion fee is £2.00

The current final value fees are:

- Item not sold, no fee.
- £0.01 to £29.99, the final value fee is 5.25% of the amount of the high bid.
- £30.00 to £599.99, the final value fee is 5.25% of the first £29.99 (£1.57), plus 3.25% of the remaining balance.
- Over £600.00, the final value fee is 5.25% of the first £29.99 (£1.57), plus 3.25% of the next £30.00 to £599.99 (£18.53) plus 1.75% of the remaining balance.

If an item is given a reserve price (which has to be a minimum of £50.00) the insertion fee is 2% of the reserve price, up to £4,999.99 and £100 for items over £5,000.00.

Users can also pay to have upgraded features, such as featuring an item at the top of a page of search results or adding bold highlighting. Additional picture options can also be purchased.

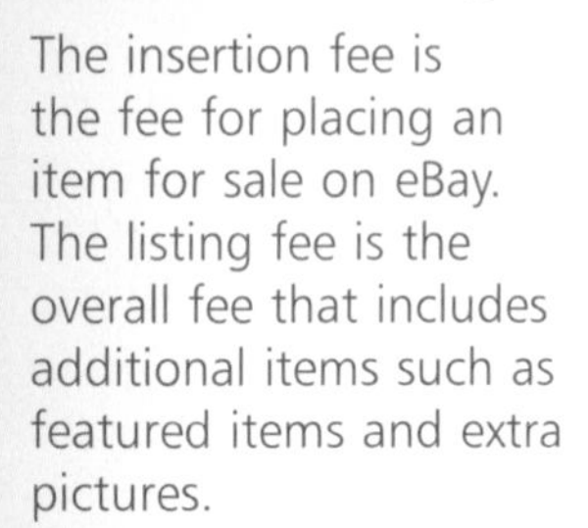

Don't forget

The insertion fee is the fee for placing an item for sale on eBay. The listing fee is the overall fee that includes additional items such as featured items and extra pictures.

Hot tip

If you are selling real estate on eBay, the insertion fee is fixed at £35.00.

Don't forget

Prices given in this book were correct at time of going to press.

Methods of payment

There are several ways in which goods can be paid for on eBay. If you are selling items, you will have the option of selecting the payment methods which you are prepared to accept. If you are a buyer you will be able to see the accepted payment methods on the auction page. The methods of payment on eBay are:

- Electronic payment. This is where an online payment is made through a third party. On eBay the most common form of electronic payment is through PayPal, which is looked at in more detail on the following pages.
- Cheque. A personal cheque can be used to pay for items. However, this takes longer as the seller should wait until the cheque has been cleared before sending the goods.
- Credit card. Some sellers accept direct payments by credit card. However, if you are a seller this can involve charges for each credit card transaction.
- Postal order or money order. This is a way of sending money without using cash or a cheque. They can be purchased at post offices or banks but there is usually a charge to cash them.
- Escrow. This is a service where a company holds payment for an item until the transaction has been completed. It is looked at in more detail on page 166.
- Bank transfer. This is where money is transferred directly from one bank account to another. There is usually some form of charge attached.

For all of these methods of payments, eBay offers some level of protection against fraudulent transactions. Under the standard purchase protection programme, payments are covered up to £120 (minus £15 processing costs).

It is strongly recommended that cash is never used as a method of payment: it can easily be stolen in transit and it is untraceable if the seller does not send the purchased item.

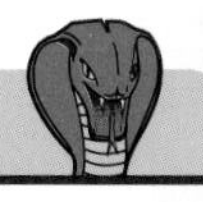

Beware

Any type of payment is subject to the possibility of fraud. If you are uncertain about any type of payment, contact the seller for clarification. If you are still in doubt, do not proceed with the transaction.

Beware

Never make payments through an instant cash wire transfer service such as Western Union or MoneyGram, particularly if this is the only method of payment offered. It means that you effectively pay cash, without any guarantee that the goods will be delivered.

Don't forget

Cash On Delivery (COD) can be an option for payment, but only if the goods are being delivered in person by the seller. This could be if they live near to the buyer and it should only really be used after contact between the two parties.

About PayPal

Don't forget

PayPal's customer service has come in for some criticism in the past, so if you do have a problem with the service it could be time-consuming to resolve it.

Don't forget

Despite some of the issues connected with PayPal, it is still a very effective and efficient way of completing transactions on eBay.

Hot tip

If you want to file a claim under the PayPal Buyer Protection service, it has to be made within 30 days of the PayPal payment.

PayPal is an online electronic payment service and it is eBay's preferred method of payment for buyers and sellers. The company was bought by eBay in 2002 and conducts billions of pounds worth of transactions every year. PayPal works by sending funds directly from your PayPal account to the seller of an item. Once you have set up an account (see next page) you can use this to send payments using either your credit card or your bank account, and you can also use it to accept payments from other eBayers.

Once you have set up a PayPal account you can select this as a method of payment when you are listing items for sale. When the auction page is published, this is denoted by the following icon:

After a sale, the money is held in your PayPal account and you can then have it transferred to your own bank account or receive the money as a cheque. You can also keep the money in your PayPal account and use this for future purchases on eBay.

Although PayPal deals with huge amounts of money it is not a bank, and so does not have to comply with the same rules and regulations as traditional financial institutions. This has caused some concerns about customer protection and there have been legal cases to try to ensure greater protection for PayPal customers against fraudulent practices. As a result there have been some improvements to PayPal and the service now offers a Buyer Protection service where qualified purchases are covered up to £500.00.

There have also been issues with chargebacks: the system whereby the buyer can claim back money from a purchase if they are not happy with the product. In some cases this has happened when the goods have been perfectly acceptable and the seller has had money removed from their account.

Registering for PayPal

To use PayPal you have to first register to set up an account. To do this:

1. Go to the PayPal home page at **www.paypal.com** and click the Sign Up Now button

Join PayPal Today
Now Over
96.2 million accounts

2. Select the type of account and click Continue

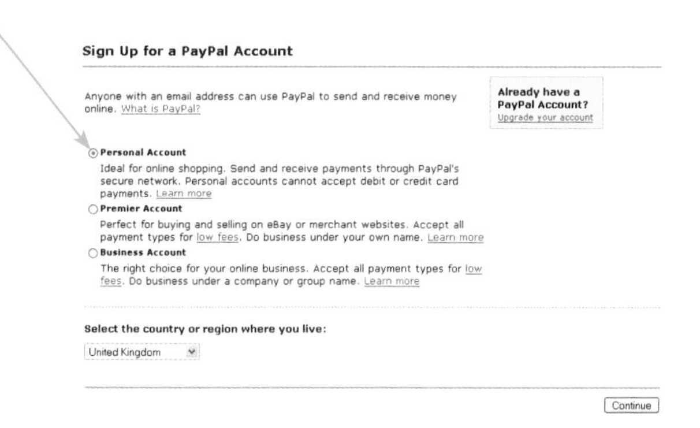

3. Enter your personal details. This will include details of your credit card, which can be used to make payment but at this stage is just used to verify your identity

4. Click this link to activate your account

PayPal
The way to send and receive money online

Activate Your Account.

Dear Nick Vandome,

Congratulations. **You've just created a PayPal account.**

But you've not finished yet.

You must **click the link below** and enter your password on the following page to confirm your email address.

Click here to activate your account

Hot tip

Make sure you read the PayPal User Agreement very carefully before you sign up for the service.

Paying with PayPal

Once you have set up and activated a PayPal account you can use it to pay for items that you have purchased on eBay, as long as the seller accepts this method of payment. You do not have to deposit any money in your PayPal account, as this will be debited from the credit card or bank account details you entered when registering for the service. To make a payment with PayPal:

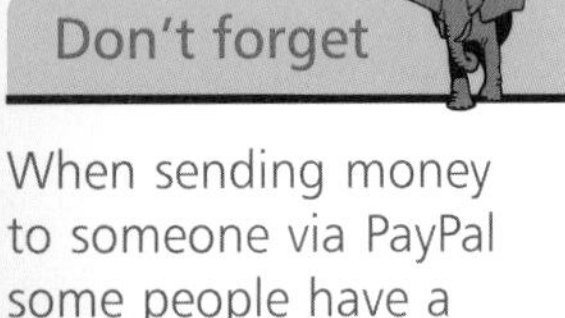

When sending money to someone via PayPal some people have a 'Confirmed' address. This involves a verification process to prove that the person really does live at the address they claim to. However, this is only available in certain countries and is not currently available in the UK. If someone does not have a Confirmed address it should not be a matter of undue concern.

1. Once you have won an auction the following message appears. Check that PayPal is accepted and click here

2. Details of the required payment are displayed

	Qty.	Price	Subtotal
	1	£0.99	£0.99
Postage and packing via Seller's Standard Rate:			£3.99
Postal insurance: (Optional £1.00)			Add
Seller discounts (-) or charges (+):			Add
Seller Total:			**£4.98**
			recalculate
Questions about the total? Request total from seller			

3. Ensure PayPal is displayed as the payment method

Confirm payment method

You are eligible for up to £500 PayPal Buyer Protection

4. Click Continue

Continue >

5. Click here if you have not already set up a PayPal account

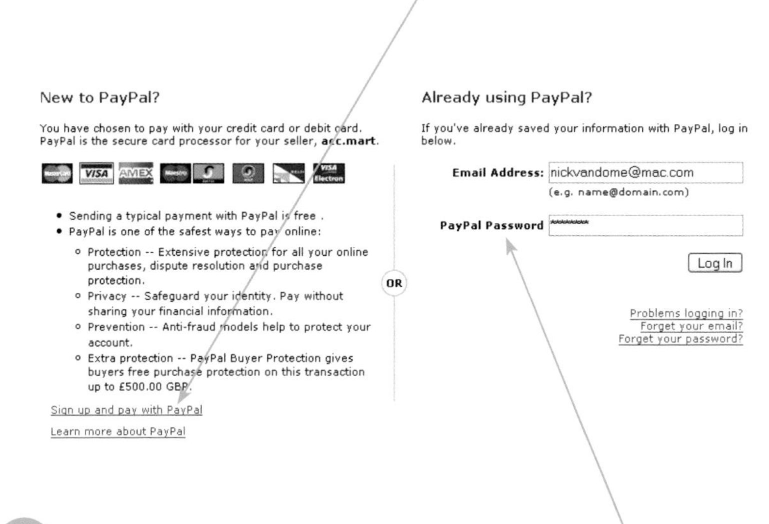

6. Enter your PayPal login details (which you specified when the account was first created) and click Log In to be taken through the full payment process on the PayPal site

7. Once the payment has been made, you will receive a confirmation email with details of the transaction

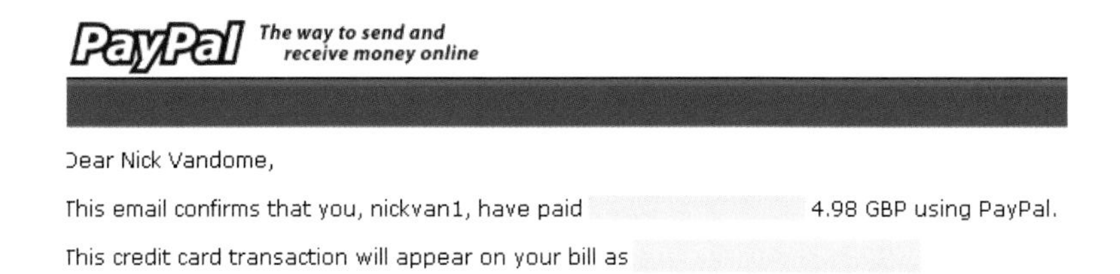

PayPal
The way to send and receive money online
Dear Nick Vandome,
This email confirms that you, nickvan1, have paid 4.98 GBP using PayPal.
This credit card transaction will appear on your bill as

Receiving PayPal payments

If you are selling an item and you have a PayPal account, you can use this to receive payment. However, in some cases you may need to upgrade your account to do so, such as if a user uses a credit card to make a payment to you via PayPal. If this happens, the payment will be marked as Pending in your PayPal account and you will have to accept or refuse it. If you accept, you will have to upgrade to a Premier account in order to obtain the payment. If you refuse, the buyer will have to send another form of payment. To accept a payment and upgrade a PayPal account:

Don't forget

For a standard PayPal account it is free to open the account and send and receive funds. However, there is a charge for withdrawing funds. This is £0.25 for £49.99 or less, but free for £50.00 or more.

1. If a payment has been made by credit card, you will be notified and asked if you want to accept or refuse the payment. Click here to accept

Accept or Refuse a Payment

Before accepting this payment, please consider the following:

Do you plan to withdraw funds from your PayPal account to your bank account?
If so, you must first add a bank account to your PayPal account before you will be able to make a withdrawal.
Add a Bank Account

Note: If you accept this payment, any other payments that are being held in your account with Pending status will automatically be accepted.

Refuse Payment | Accept Payment

Don't forget

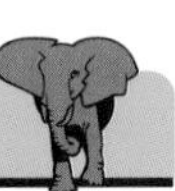

The fees for a Premier account are the same as for a Standard one, except that there is also a charge for receiving funds. This is 1.9% plus £0.20 up to 3.4% plus £0.20, depending on the amount received. This obviously has an impact on anyone who is receiving credit card payments. It is free to upgrade to a Premier account.

2. If you have a standard account you will be taken to a window for upgrading your account, which has to be done in order to obtain the credit card payment. Click here to upgrade your account

Upgrade Your Account

Premier Accounts include all the benefits of a Personal account PLUS the following premium features:

- Accept **credit card payments**
- Accept payments on your own website or personal homepage
- Set up your own PayPal Shopping Basket
- Exclusive customer service hotline seven days a week
- Identify yourself as a Premier account in your reputation pop-up box, visible to anyone paying you

Get all of our premium features for a low fee on incoming payments.
See Fee Schedule

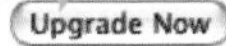

Buyer protection

All buyers on eBay like to know that they are operating in as secure an environment as possible. However, some fraudulent activity does take place and eBay tries to boost user confidence with its Standard Purchase Protection policy. This is offered for most auctions and it is denoted by this symbol on the auction page:

Standard Purchase Protection Offered.
Find out more

Standard Purchase Protection provides the buyer with insurance against being subject to a fraudulent transaction, for example sending money but not receiving the goods. The maximum amount of compensation that they can receive is £120, minus a £15 administration fee. So, if the item was £200, the buyer would receive only £105. If the item was £50 the buyer would receive £35: the amount of the claim minus the £15. There are certain criteria that have to be met for this:

- The item was listed and purchased on eBay.
- Both the buyer and seller have positive feedback ratings.
- The item complies with eBay's User Agreement and policies.
- The claim is filed within 90 days of the auction closing.
- You paid for an item and did not receive it, or it was significantly different from the described item.
- No more than 3 claims per 6 months for each user.
- The item was not paid for by cash or an instant cash transfer.
- The items were not lost or damaged in transit.
- The service is not available just because the buyer does not like an item or changes his or her mind.

Hot tip

If you want additional protection when selling expensive items then an escrow service should be considered, as the Standard Purchase Protection would not offer a high enough level of protection. See the next page for details.

Escrow

Escrow is a type of service that tries to minimise the chance of fraud during an eBay transaction. It works by holding the money until the sale has been completed to the satisfaction of both parties. In general there are five steps in a transaction that makes use of escrow:

1. The buyer and seller agree to use escrow and agree conditions such as the item for sale, its price, the time the buyer has to inspect the item, and the shipping details.
2. The buyer pays the escrow service.
3. The seller sends the item.
4. The buyer inspects the item and either accepts or rejects it.
5. The escrow service pays the seller (if the buyer accepted the item).

There are a number of online companies that provide escrow services and the one recommended by eBay is:

Escrow.com at **www.escrow.com**

For users in the UK and the rest of Europe, the following escrow service can be used: **www.escrow-europa.com**

Most escrow services charge fees based on the value of the item being sold. For transactions in GBP the fees on **escrow-europa.com** are as follows:

- Up to £3,500: 2% of the transaction
- Between £3,500.01 and £17,000: 1.5% of the transaction
- Between £17,000.01 and £176,000: 1% of the transaction
- Between £176,000.01 and £352,000: 0.85% of the transaction
- Over £352,000.01: 0.65% of the transaction

Don't forget

Escrow is a generic term rather than one specific service, even though the eBay-recommended escrow service goes by that name.

Don't forget

Escrow is generally used for expensive items, where there is a considerable risk for both the buyer and the seller.

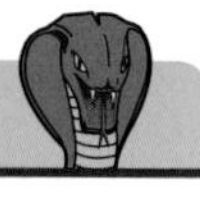

Beware

There are some fraudulent escrow services that try to entice people into lodging money with them. If you want to use an escrow service then stick to escrow-europa.com, or escrow.com for business outside Europe.

11 Business on eBay

Thousands of people make a living solely from buying and selling on eBay. There are fewer overheads than running a 'bricks and mortar' retail business.

Preparing for business

While many people simply use eBay as a means to clear out their attics and garages, thousands more have created online businesses exclusively on the site. It is possible to make a living from trading on eBay, but if you are thinking of this then you have to approach the whole venture in a professional manner. There are certain areas that you should look at before you start your business:

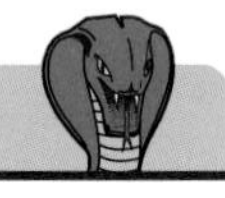

Beware

Just because you sell a handful of items on eBay, this does not necessarily mean you will be able to run a successful business. Examine all of the angles very carefully and consider all of the things that could go wrong.

Don't forget

In theory, all income generated on eBay is subject to income tax. This is particularly relevant if you are operating as a business. However, if you are only selling a few items as a hobby and for a small amount of money, you may not have to worry about tax. If in doubt, check with an accountant.

Business plan

An online eBay business is the same as any other as far as having a business plan is concerned. This is not only essential for any institution from which you want to borrow money, it is also invaluable for getting clear in your own mind what you want to do. Some areas to consider when you are creating a business plan are:

- The type of product or products you are going to sell and how you are going to source them.
- The money required for the initial investment, which could include the products to sell, office space and warehouse space.
- The amount of time you are able to commit to the eBay business (a lot of eBay businesses are run by people who are also working in another job).
- Estimated income and expenditure, in both the short-term and long-term. Make sure that these estimates are realistic.

Several business enterprise organisations or local councils offer advice about setting up a small business, including details about creating a business plan.

Initial investment

The main reason for creating a business plan is to borrow money from a financial institution. Their chief concern will be whether they will be able to get back any investment which they make. As a potential eBay business person, your first task is to decide how much money you will need to set up and run your business. In some cases you may be able to finance this yourself (at least initially) or to persuade family or friends to invest in your business.

If you do go to a bank or a building society for a loan, arrange a personal meeting so that you can discuss your idea and business plan face-to-face with someone. Do not over-extend yourself in terms of the amount of money you borrow, and make sure you ask questions of the financial institution, such as how much they will charge for the loan and the fees that will be charged on a business bank account (if you set up one). In some cases, the lender may ask for some form of collateral, but try and avoid this by limiting the amount you want to borrow.

Beware

Always be careful when borrowing money from family and friends: what might seem like a great idea in the excitement of a new venture can turn sour once the realities of the business world start to bite.

Being professional on eBay

If you are looking to operate an eBay business then you have to do this in as professional a manner as possible: if you are looking to make it more than a hobby then you have to treat it as such. This includes:

- Making your listings as professional as possible in terms of product descriptions and accompanying photographs. One way of making your listings more business-like is to create your own eBay Shop, and this is looked at on the following pages.
- Decide how you want to set up your business in terms of being a sole trader, a partnership or a limited company. This will depend on the size of the prospective business and the number of people involved.
- Find a lawyer who is a specialist in business law. This will not only help you if there are any disputes once you start your business, but they will also be able to give you advice about setting up the business.
- Find an accountant. If you are serious about trading commercially on eBay then it is important to have someone to do the financial books and deal with thorny issues such as tax and VAT.

Don't forget

Although it costs money to employ lawyers and accountants this could save you money in the long term. A lawyer could help resolve financial disputes and an accountant could make you aware of expenses for which you can claim. Also, fees for professional services such as these are tax-deductible as business expenses.

Inventory

The inventory for an eBay business is basically the items that are going to be sold. These can be sourced from the same areas as for those for individual sales. However, if you are setting up as a business you will probably want a larger volume of items than for an individual seller. Some areas for obtaining inventory include:

- Manufacturing items yourself. If you have a talent for producing something then you may be able to create a niche market for it on eBay. However, bear in mind that this will eat into your time for running and managing the online side of your eBay business. A good compromise is to find local people who produce exclusive products and then discuss the possibility of setting up a joint business venture.
- Wholesalers. As mentioned on page 92, buying goods in bulk from wholesalers is a good way to obtain a large amount of inventory at reasonable prices.
- Trade shows. Almost every trade is represented, at some point, at a trade show. These are large meeting places for specific trades and they allow companies to show off their products and to sell them to prospective buyers. It is a great way to meet people, to see what is on offer and to make contacts. For a comprehensive list of trade shows in the UK, look at **www.exhibitions.co.uk**
- Trade associations. Most trades have equivalent professional associations. If you are interested in a particular trade it could be worth joining its association for contacts and ideas.
- Closeouts, clearances and stocklots. There is a whole industry of companies who buy surplus stock at cheap prices from a variety of retailers. There are a number of reasons for this, such as liquidations, stock that has not sold before a new season starts, or unclaimed freight. A list of some of these companies can be found by typing 'closeouts +uk' into Google. It is also possible to do this type of business yourself with local retailers.

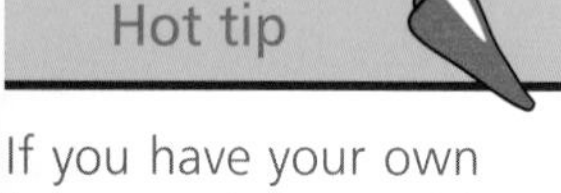

If you have your own logo, or photographic image, you can add this to your shop front by selecting this option and browsing to the file on your hard drive.

Hot tip

Some companies use inventory software to post items on eBay shortly before they reach the end of their shelf lives. This can be a good way of moving unwanted stock before it is too late. One such program (from the German company DeNISys) can be incorporated into the popular business-management software SAP Business One.

Opening an eBay shop

Once you have decided on the type of business you want to run and what you will be selling, it is time to think how to present your stock to potential customers. One way to do it is to run numerous individual auctions, but this can be hard to keep track of. A more professional approach is to open your own eBay Shop. This will then give you a single, business-oriented location for your business. To set up an eBay shop:

The eBay User Agreement for opening a shop is the same as the one for trading in individual auctions on the site.

1. Login and click the eBay Shops link on the home page

:::Speciality Sites
Business Centre
eBay for Charity
eBay Shops
PayPal
Reviews & Guides

2. Click the Open a Shop button to open a new shop

Open a Shop
Manage My Shop

3. Click on this button

Open Your eBay Shop >

4. Select a design for your shop home page

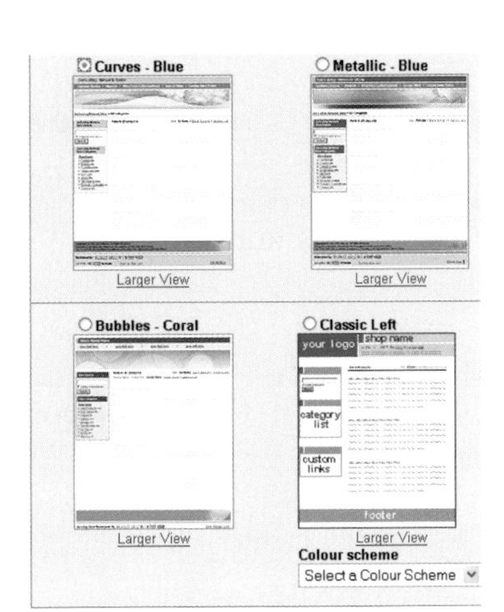

5. Click Continue

6 Give the shop a name

Shop name

Nick's Shop

24 characters left.

Hot tip

If you have your own logo, or photographic image, you can add this to your shop front by selecting this option and browsing to the file on your hard drive.

7 Select a logo and click Continue

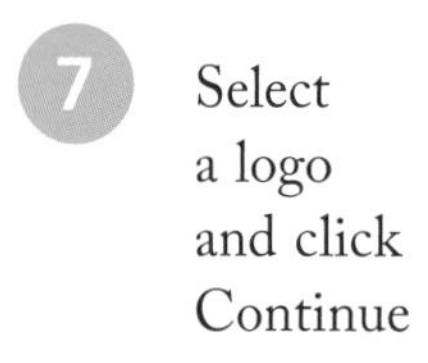

Continue >

Shop Logo
Graphic size is 310 x 90 pixels. Other sizes will be automatically resized to fit these dimensions. Learn more about including your logo
Use a pre-designed logo:
Collectables
Computers
Dolls & Bears
Electronics
Hobbies & Crafts
Upload a logo to Picture Manager
Picture Manager is a service offered to Shop subscribers by eBay
Browse...
Preview
No Logo

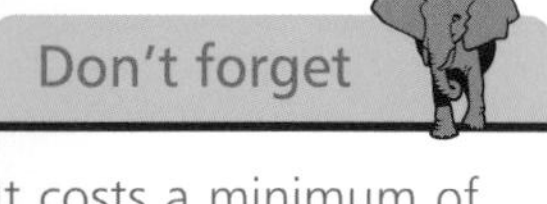

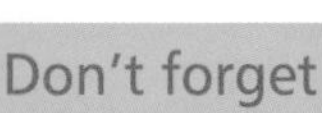

Don't forget

It costs a minimum of £6 per month to have an eBay shop, but this is a small price to pay considering the potential advantages. If you want your shop to be more actively promoted it is more expensive.

8 Select the type of shop which you want to create. The default is Basic Shop

Subscription Level

Choose a subscription level that best mee
Shop must comply with eBay policies. eE

Basic Shop (£6.00/month)
A Basic Shop is an ideal solution fo
online.

Featured Shop (£30.00/month)
A Featured Shop offers all of the ber
small to medium-sized sellers who v

9 Click Start My Subscription Now to create your shop

Start My Subscription Now >

Auction management software

If you have only ever participated in a few individual auctions you may think that it is easy to keep track of your items, bids, questions and payment options. However, if you are running a commercial venture on eBay you may find that you have dozens, or even hundreds, of auctions running simultaneously. This can lead to stress and confusion as you try to keep up with everything that is going on with your auctions. In a worst-case scenario your business could collapse under the sheer volume of transactions. This is where auction management software can help in automating as much of your eBay activities as possible. Some of the areas which auction management software can handle are:

- Stock control. This keeps a track of what you currently have for sale and what has been sold. As you add more stock, this is updated within the auction management software.
- Auction page creation. Creating standarised auction pages from templates is a great timesaving device and helps achieve consistency across your auctions.
- Auction management. This keeps a track of your auctions so that you can see what is going on.
- Communication. This is a vital part of eBay and auction management software can take care of filtering your eBay questions, generating automated acknowledgments and sending notification to winning bidders.
- Payment. This is another area that can be largely automated by the software. As soon as an auction finishes, automated receipts can be sent to the winning bidder. It is also possible to set up online payment facilities and create shipping details.

eBay has its own auction management software, called Turbo Lister. There are also many commercial programs available; one of the market leaders is called Andale. It offers a variety of auction management solutions – more information can be found on the website at **www.andale.com**

Don't forget

The My eBay page is a very basic form of auction management software, as it enables you to see details about different auctions.

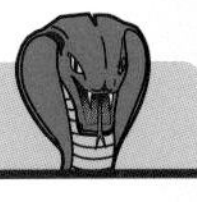

Beware

Auction management software can automate only some of your communication with customers. Sometimes you will have to contact them directly, usually by email. Do not forget the human factor just because the software can do so much.

Hot tip

Turbo Lister is currently free to download (although it is only available for use with Windows). More details can be found at **http://pages.ebay.co.uk/turbo_lister**. Another similar eBay program is Selling Manager.

Trading for others

As well as selling your own goods on eBay it is also possible to sell items for other people. People who sell items on behalf of others are known as Trading Assistants. If you want to set yourself up as a Trading Assistant you have to meet the following requirements:

- You must have sold at least 4 items in the last 30 days, to show that you are an active trader on eBay.
- You must have an overall feedback score of 50 or higher.
- You must have an overall positive feedback rating of 97% or higher.
- Your eBay account must be in good standing, with no ongoing disputes or investigations.

If you meet these requirements you can register for the Trading Assistants Directory. This can be done as follows:

1. Click the More Tools & Programmes link on the home page

See Also
a/A Change text size
Answer Centre
Community Hub
eBay Groups
Gallery
More Tools & Programmes
Seller Tools
Want It Now

2. Scroll down the page and click here

Trading Assistants 

Trading Assistants are experienced eBay sellers who are willing to sell items via eBay on your behalf for a fee. You can find an eBay Trading Assistant who lives close to you by entering your post code or county and searching.

Show me more...

3. Click here to create an entry in the Trading Assistants Directory

Choose a Topic
Find an Assistant
How It Works
FAQs
For Trading Assistants
Create/Edit Your Profile
Getting Started
FAQs
Toolkit
Discussion Boards

Don't forget 

Trading Assistants usually trade for themselves as well as for others. This means they frequently have a high turnover of sales and are often Power Sellers.

Don't forget 

To find a Trading Assistant, follow Steps 1 and 2 on this page. Enter your post code to show all of the Trading Assistants in your area.

When you register for the Trading Assistants Directory you will have to give details of your service, so prospective clients can see what you offer. This information can include:

- How you will obtain the items from the sellers: whether you are prepared to pick things up or if you have a drop-off point where the sellers can take the goods.
- How much you will charge for selling items. This is usually a percentage of the final selling price.
- Whether the eBay fees are paid up front by the seller or included in the Trading Assistant's fee.
- Types of items which you are prepared to sell. Most Trading Assistants have particular areas of expertise.
- How the item is shipped once it is sold.
- How you are going to pay the seller once the item sells.
- Return and relisting policy (if the item does not sell in the first auction).

Operating as a Trading Assistant

Once you have registered in the Trading Assistants Directory you will be able to start selling for other people. This is done by the seller contacting you via the Trading Assistants Directory and then agreeing either to drop off the item or for you to pick it up. In general it is only really worthwhile if the expected selling price for the item is £50 or over. Once you have the item, you list it on eBay and conduct the auction in the usual way.

As a Trading Assistant you will be subject to the same auction conditions and feedback criteria as if you were conducting an auction for yourself. Although being a Trading Assistant can be more work than running your own auctions, it is an excellent way to supplement your eBay income, particularly if it is used in conjunction with your own eBay shop.

Hot tip

When entering your details for the Trading Assistants Directory, be as comprehensive as possible. This is your sales pitch for people who are looking for a Trading Assistant, so ensure you make as compelling a case as possible for them to use your service.

Storage and office space

If you are serious about running a retail business on eBay then you will have to give some consideration to how you are going to store items for sale and also where you are going to operate your business from.

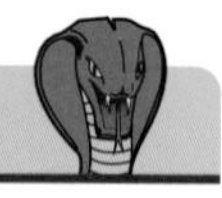

Beware

It is not a good idea to have too much stock being held in storage at any one time. Try to rotate it as much as possible and avoid creating a backlog of items to sell.

Storage

Initially, you may be able to keep the items you hope to sell on eBay in a spare room or a garage (depending on their size). However, if you are constantly looking to expand your inventory, these options may quickly become unrealistic. If your current storage options start to become cramped, then this could effect your overall operation. Items could be lost or misplaced, resulting in unfulfilled orders and dissatisfied customers. If you are looking for additional storage some of the options could be:

- A lock-up garage. These can be rented for monthly or annual fees and provide a good-sized storage facility.
- Commercial storage services. There are companies that offer a variety of storage options such as units, or containers within a warehouse, which can be rented short or long term. Check your local Yellow Pages or search on the Internet for details.
- An additional shed at home. Garden sheds can be a relatively cheap way to expand your storage, but make sure you have enough room and that they are secure.

Don't forget

Storage space can also be used to package items before they are sent to customers. However, make sure there is extra room for this with items such as a table and packaging materials close at hand.

Office space

It is possible to operate an eBay business from the kitchen table, but it is a lot more efficient, and satisfying for all concerned, if you have some dedicated office space. This can be hired commercially but a more realistic option could be to use a spare room and convert it into an office. It does not have to be huge, but make sure it has good lighting, sockets for telephone and computer equipment and adequate shelving for books and paperwork. Also, make sure anyone else living in the house is aware that the room is a work office and not a playroom or an extra TV room.

Thinking globally

eBay really is a global marketplace and if you are operating a business then you should be looking further afield than the markets within your own country. It is perfectly possible to sell your goods anywhere in the world, but there are some points to consider before you start branching out beyond your own shores:

- Local restrictions. Global eBay sites can have different rules concerning what can and cannot be traded, depending on the laws of the country in which they are situated. So it is important to not only comply with the regulations within your own country but also those for the country in which the buyer is located. Make sure you read the user agreements for any countries where you will be doing business.
- Export and import issues. Make sure that you comply with the export conditions for your own country and also the import conditions for the country where the goods are being sent. Your items must clear customs when they are entering a country, so you will have to complete the relevant paperwork when sending items. This could involve a description of the item, where it is being sent and its value. If the item does not clear customs it will be returned to you.
- Communication. When you are selling abroad you may be dealing with customers whose first language is different from yours. If you are in doubt about anything, get it translated by either an online translation service or an actual translator.
- Payment. The best method for international payment is PayPal, but remember to check the exchange rate for the country involved, to see how much you will receive. Also, PayPal charges a fee for converting between currencies.
- Shipping. Use a recognised mailing service such as the Royal Mail or a courier service, and check that they deliver abroad. It is worth getting postal insurance, but make sure you include this in the overall postage charge.

Hot tip

Include exact details about where in the world you are prepared to post items if you are operating globally.

Hot tip

There are a lot of online translation services that will translate text for free. One to try is **www.freetranslation.com**

Customer service

Some of the most successful retail companies are the ones with the best customer service. If you are operating your own business on eBay try and emulate this and become known for first-class customer service. The reason for good customer service is not just to ensure that first-time customers are happy with their purchases: the real benefit is in getting these customers to keep returning. Good customer service should be at the top of your list of priorities. Some points to consider are:

- Make sure you give people what you say you will. The first step in good customer service is to ensure that what you describe and display in an auction is exactly what the buyer receives. There is no point in trying to hide defects or to exaggerate certain features. These embellishments will be obvious once the buyer gets the item, and then you will have an uphill battle to convince them to buy from you again.
- Communicate quickly and effectively. Answer questions from customers as quickly as possible, so that they can see you are professional and efficient. This will help build up their confidence when dealing with you, which is a vital part of customer service.
- Remember the personal touch. Be friendly when you contact people: use their first names in emails and include a compliment slip when you post items.
- Go the extra mile. Try to accommodate requests from buyers or prospective buyers. Even if it is something that may not immediately gain a sale or make extra money, it could pay off in the longer term if the person comes back to buy something else from you.
- Accept unconditional returns. If you state that you accept returns without any questions asked, buyers will be even more confident about dealing with you. A lot of successful high-street stores have this type of return policy.

Don't forget 

When it comes to communicating with buyers, keep calm regardless of any negative or provocative comments they may make. If necessary, it can be better to cut your losses in a dignified manner rather than getting into a protracted slanging match, which could eventually translate into negative feedback.

12 Troubleshooting

Things can go wrong with auctions and with interactions between users. Here we look at some of these problems and how to avoid them or deal with them.

Dispute resolution

Due to the number of transactions that take place on eBay it is inevitable that disputes will occur between buyers and sellers. This could be because of defective or fake goods, non-payment, or non-delivery of items. If this happens, buyers can try eBay's own buyer protection service, as detailed in Chapter Ten. Another option, which is recommended by eBay, is to use a dispute resolution service. Their preferred option is Square Trade, which can be found at **www.squaretrade.com**. The service operates as follows:

- The first party files a complaint against the second party.
- The second party is contacted by Square Trade and told the nature of the dispute.
- The two parties are invited to discuss the dispute within a secure page on the Square Trade site. This is known as 'direct negotiation' and there is an option for a Square Trade mediator to help resolve the issue.
- Ideally, the case is resolved to the satisfaction of both parties

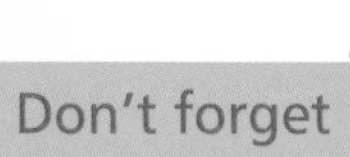

Before using a dispute resolution service such as Square Trade, try to resolve any problems directly with the other party involved.

To use the Square Trade service:

1. Visit the website at **www.squaretrade.com**

2. Click File a Case to start a new dispute resolution case

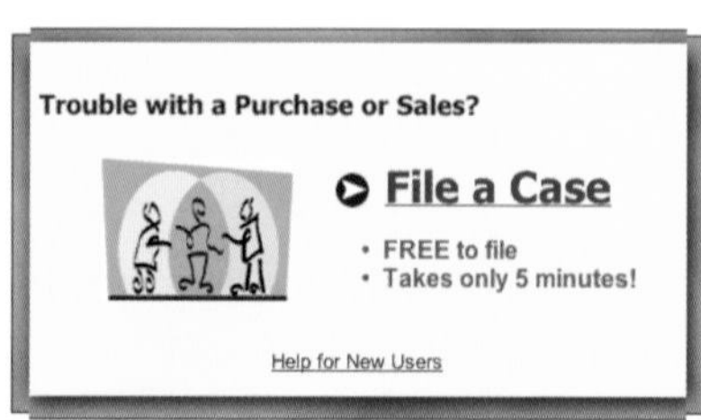

Trouble with a Purchase or Sales?
File a Case
• FREE to file
• Takes only 5 minutes!
Help for New Users

3. If a seller displays this symbol, it shows they have met certain Square Trade criteria for fair trading

Scams and frauds

In any walk of life, large sums of money will attract criminals and con artists. eBay is no different in this respect and some individuals undertake fraudulent activities on the site. However, that said, fraud forms only a tiny percentage of the overall volume of sales, and you are probably as secure trading on eBay as you are in any other retailing environment. The type of frauds that can occur on eBay are:

- Non-delivery of goods. This is the most obvious fraud, and the one that users (particularly new users) are most worried about. This is a relatively simple fraud, whereby the seller takes payment for an item and never sends it. Thankfully, due to the robust feedback system on eBay, these individuals do not last long on the site. However, they can have time to make a number of fraudulent transactions before they are caught out. As a buyer all you can do is try to investigate sellers as fully as possible from their feedback before you make a payment. If you are in any doubt about the type or amount of feedback they have then do not proceed.
- Long-term fraud. Some individuals have the patience to spend months building up positive feedback and a good trading background on eBay. Then they start taking payment and not sending the goods. Since the feedback may not offer any clue to the buyer, try looking at the types of items that are for sale. If someone has been selling low priced items and then they start selling much more expensive ones, this could be a warning sign.
- Selling links rather than products. Some auctions for items look like they are for the actual item, but they turn out to be for a link to a website that may, or may not, offer the item at a bargain price. So essentially you are paying for the possibility of buying the item cheaply. There are a number of these types of auctions but personally I would never participate in them. Having said that, a lot of the sellers of these items have high-scoring feedback which suggests they could be genuine, but in general it is better to be safe than sorry.

Beware

Legally, fraud on eBay is just the same as in any other walk of life, and people who commit fraud are liable to criminal prosecution.

Don't forget

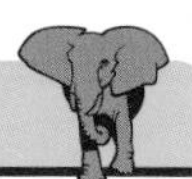

If you are subjected to a fraud, report it in the first instance to eBay via the buyer protection service. Also, depending on the amount involved, you may want to consider reporting it to the police and Trading Standards.

Spoof emails and websites

Another fraudulent practice directed at eBay is the proliferation of spoof, or fake, emails and websites in recent years. These try to get you to part with your money as part of pyramid selling schemes or pretend to be genuine communications from eBay or PayPal and then ask for your personal login details.

The distribution of fake emails and websites is known as 'phishing'.

If you have an eBay account and a PayPal account then you will almost certainly receive fake emails sooner or later. Their content will quickly give you an idea of whether they are real or not; if they ask you for money or redirect you to a website that tries to obtain your personal details then they are probably fake. The golden rules about these types of emails are:

- Never part with any money on the promise of making a fortune from pyramid selling.
- Never disclose either your eBay or PayPal login details.
- If in doubt about a spoof email, report it to eBay by forwarding it to **spoof@ebay.co.uk**

Fake websites can look convincing in terms of their appearance and the use of eBay and PayPal logos. However, the clues are usually within the content of the pages.

Even if emails or websites claim to be genuine and look legitimate, it is almost certain that they are not real if they ask for money or personal details. If in doubt, look at the web address of websites and also view the properties of emails to see if there are any obvious clues there.

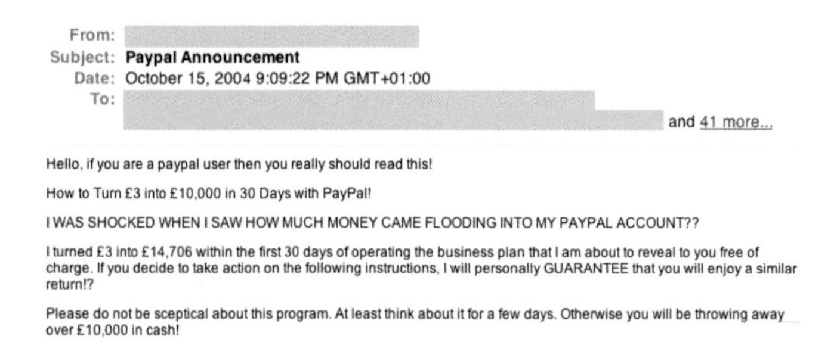

From:
Subject: **Paypal Announcement**
Date: October 15, 2004 9:09:22 PM GMT+01:00
To:
and 41 more...

Hello, if you are a paypal user then you really should read this!

How to Turn £3 into £10,000 in 30 Days with PayPal!

I WAS SHOCKED WHEN I SAW HOW MUCH MONEY CAME FLOODING INTO MY PAYPAL ACCOUNT??

I turned £3 into £14,706 within the first 30 days of operating the business plan that I am about to reveal to you free of charge. If you decide to take action on the following instructions, I will personally GUARANTEE that you will enjoy a similar return!?

Please do not be sceptical about this program. At least think about it for a few days. Otherwise you will be throwing away over £10,000 in cash!

Spoof emails should always be ignored, regardless of their content.

Account Guard

One way to increase your security against spoof websites and emails is to use eBay's own Account Guard, which is included in the eBay toolbar. This is a downloadable toolbar that can be used to keep track of your eBay activities. Account Guard is now part of the toolbar, and has a colour-coded system to notify you about the status of the website you are visiting:

- Green indicates that you are on a safe eBay or PayPal site.
- Grey indicates that you are probably on a genuine eBay or PayPal site but it might be a fraudulent or spoof site.
- Red indicates that you are probably on a fraudulent site.

At present, the eBay toolbar and Account Guard are only available for Windows. To obtain these:

1. Click here at the bottom of the home page

2. Click here to download the eBay toolbar, which has Account Guard built into it

Download the eBay Toolbar Today

3. The presence of Account Guard is denoted by this symbol

Don't forget

The eBay toolbar can be used to search for items within eBay, to receive information about ongoing auctions and also to protect your password.

Non-payment

One of the unavoidable problems when conducting an auction is non-payment. This occurs when someone places a high bid in an auction, wins the auction and then simply does not pay. The common eBay terminology for this type of behaviour is a 'dead-beat bidder'. The essential thing to remember about non-payment is to make sure you never send any goods until you have received payment. If you are paid by cheque, make sure the funds have cleared fully before you send the item. If you come across a dead-beat bidder you should try to obtain the funds as follows:

Beware
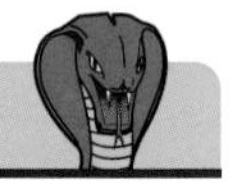

If a user has three Unpaid Item strikes against them they will be barred from eBay.

- Contact the buyer directly. Wait a few days and then email with a polite, but firm, request for payment. This should be done at least 3 days after the auction, and within 30 days.
- If there is no response to the initial contact, you can file an Unpaid Item dispute. You have to wait at least 7 days after an auction has closed, and within 45 days.
- Once an Unpaid Item dispute has been filed, eBay contacts the buyer and requests payment or asks for an explanation about the ongoing non-payment. The buyer then has the chance to pay for the item.
- If the item is still unpaid for, the buyer receives an Unpaid Item strike against their name and you become eligible to relist the item. If this happens, you receive a credit for the Final Value fee from the initial auction and can relist the item for sale without having to pay the listing fee again.

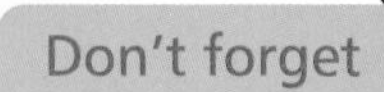

An Unpaid Item dispute can only be open for a total of 60 days.

Second Chance offers

If you have been subjected to a non-paying winning bidder you can offer the item to any of the under-bidders in the auction. This is known as a Second Chance offer. As long as the auction has closed and there is at least one under-bidder then there will be a link to activate this on the auction page and also on the My eBay page. The only fee for a Second Chance offer is the normal Final Value fee.

Inappropriate bidding

The process of bidding in an auction is not exempt from dodges and scams, and there are a number of ways in which users can try to circumvent the normal bidding protocols:

- Unwanted bids. It is perfectly acceptable for a seller to say that they do not want a bid from a particular person. This could be because the bidder has a lot of negative feedback or are bidding from a country to which they are not prepared to send the item. In this case the seller can ask the buyer to withdraw the bid. If this does not work the seller can report them to the eBay investigations team.
- Proxy bidding. This is a tactic where a buyer gets another eBay user to bid on their auction with the intention of pushing up the price. It can be hard to prove, but this type of bidding is not allowed by eBay.
- Shill bidding by the seller. Shill bidding refers to the creation of a secondary User ID, which can then be used by the seller to bid on their own auction, thus pushing up the price in the same way as proxy bidding.
- Shill bidding by the buyer. This occurs when a buyer uses a secondary User ID to place an unrealistically high bid for an item. The intention is that other bidders will see this bid and so be put off the auction. The buyer then uses their other User ID to make a lower bid that becomes the second-highest bid. Then, just before the auction finishes, the initial bid is retracted and the user gets a bargain with their second, lower bid. Although this seems like a good way to win auctions cheaply, it is absolutely against the eBay rules and regulations.
- Vigilante bidding. A lot of eBayers defend its community spirit; this includes trying to sabotage auctions that they see as morally questionable. Targets include ticket sales for large events where someone has bought tickets with a view to making a profit on eBay. Vigilante bidders use additional User IDs to place very high bids for these items and not paying once they win.

Don't forget

If you want to report a user to the eBay Investigations Team, type 'investigations' into the search box on the Help Centre pages of the Community site. This contains details of the various types of investigation which can be done by eBay.

Don't forget

You can legitimately set up more than one User ID. This could be because you want to sell separate types of products and would rather have a different User ID for each one. However, you cannot use these User IDs to bid on your own auctions.

No Longer a Registered User

If you break the eBay rules and regulations you could be banned from eBay. If this happens you will be classified as 'No Longer a Registered User' (or NARU'd in eBay terminology). If this happens you will not be able to take part in auctions as either a buyer or a seller. Some offences result in instant expulsion, while others operate on a "three strikes and you're out" basis. Some of the offences for which you may be NARU'd on are:

- Non-payment for items. If you are reported three times for this you will be NARU'd.
- Theft – receiving payment and not sending the goods.
- Account theft, whereby someone steals the account details of another user and pretends to trade as them.
- Selling prohibited items.
- Trying to sell to a buyer directly by bypassing the auction – contacting them outside the eBay framework and offering to sell them an item that appears in an auction.
- Sending spam emails to eBay members.
- Sending abusive or threatening emails.
- Revealing contact details of eBay members.
- Feedback extortion – threatening to leave negative feedback for another user unless they co-operate with you in some way.
- Providing false information when you register for a User ID. This can include telephone numbers and email addresses.
- Using images from other auction pages.
- Non-payment of eBay fees.

Don't forget

It is possible for users who are NARU'd to re-register under a new User ID. However, when they do this, they will lose any feedback from their previous User ID.

Beware

Never reveal your User ID or password in reply to an email that has been sent to you. eBay never asks for this type of information in an email; it is a method used by fraudsters to take control of your User ID and your eBay account.

Index

A

B

C

D

E

F

M

N

O

P

Q

R

S

T

U

V

W